The Science of Selling Alarm Systems

WELCOME
JOHN DEVANEY

The Science of Selling Alarm Systems

Norman C. Eisenstat

Illustrations by John Devaney

Butterworth–Heinemann
Boston London Singapore Sydney Toronto Wellington

Eisenstat, Norman C.
The science of selling alarm systems.

Includes index.
1. Selling—Security systems. I. Title.
HF5439.S39E37 1984 621.389'2 84–9423
ISBN 0–7506–9174–3

Butterworth-Heinemann
313 Washington Street
Newton, MA 02158-1626

10 9 8 7

Printed in the United States of America

For my sons,
Philip and Adam.
They make it all worthwhile.

Contents

Money is a symbol of duty; it
is a sacrament of having
done for mankind that which
mankind wanted or needed.

Samuel Butler

JOHN DEVANEY

Preface

This book was written *for* and *about* money! It's also *for* and *about* helping people. But, of course, they're really one and the same, aren't they? Because when we help people fulfill their needs or wants, or when we help them to make positive decisions, we are generally rewarded with varying amounts of money (if we're in the business world). As Henry Ford answered, when asked to put into one sentence the formula for acquiring wealth, "Find a need, and fill it."

For those of us who have been either fortunate enough or wise enough to become a part of the security business, the possibilities for acquiring wealth are breathtaking. We're in a business that delivers products and services for which there is a critical need—a need that, for the most part, has gone unfulfilled.

Not only are the growth and profit potentials in the security business enormous, but at the same time, security *professionals* are able to enjoy the satisfaction of knowing that every home in which a security system is installed is one family that's not a candidate for becoming the victim of burglary, fire, or worse.

Successful security systems salespeople can generate not only impressive financial rewards for themselves but can also take pride in the fact that they are selling products and services that save lives and protect property. It's really like performing a public service while getting paid like a top executive. This book is about "how to get there."

Your purchase of this book has made a meaningful contribution to my personal income, but I can guarantee that if you learn the concepts and implement the techniques detailed here, it will make a much bigger contribution to *your* personal income, one that will be ongoing for the rest of your life.

The Science of Selling Alarm Systems

1

Alarm Industry Negatives Can Make You Rich If You Become a *Pro*

You Can't Get a Rainbow Without a Little Rain

Here's the secret of how to get very comfortable financially from the alarm industry: Take all of the apparent negatives in the business and deal with them as wealth-producing opportunities.

Sound simple? Sound easy? It's neither! It's hard—exceedingly hard. But the rewards, should you decide to make the necessary commitment, can be beyond the most optimistic dreams of most men and women.

Exactly what is it that you're going to have to commit to? You must commit to becoming a skilled *professional* salesperson, as opposed to being an amateur peddlar. Throughout the remainder of this book we'll detail just exactly what you'll have to learn and what you'll have to do to accomplish this.

Right now, the alarm industry is "hot." But, in spite of the fact that the security business is one of the country's fastest growing industries, the surface has barely been scratched, in terms of the degree to which the market has been penetrated. Think about this for a moment: Less than 2% of the 81 million residential dwelling units in the nation have any kind of security system. That's right, 98% of the homes in the United States do not have what we in the security business are selling. And, while a considerably higher percentage of the nation's business establishments have some kind of security system, the fact remains that many more businesses do *not* have a security system than *do* have one.

If we were laymen who knew nothing about either security systems or professional selling methodology, we would then surmise that selling security systems should be just like selling dollar bills for fifty cents! This is especially true in light of the fact that almost nobody has what we're selling, and the crime problem, which ostensibly dictates the need for security systems and creates the market climate, has never been more widespread and dangerous.

Paradoxically, while a veritable fire storm of crime is sweeping the country, and while burglary is approaching not only epidemic, but also pandemic proportions, the security industry has been almost totally incapable of making a significant penetration into the potential market—especially the residential market.

In the 1980s, the sciences of marketing and sales have evolved to impressive levels of sophistication and effectiveness. Marketers and sellers of consumer products, both packaged goods and durables, have, through innovation and imagination, managed to generate previously unthinkable amounts of sales dollar volume by creating products and services that capitalize on cultural, sociological, and demographic trends. Fast food, personal computer, video game, and dozens of other industries have soared to breathtaking heights because they have accurate insights into their potential markets. They simply created on-target products and/or services and then created or utilized channels of distribution that made sense.

The security industry, in contrast, has been almost medieval in its sales and marketing approach. In fact, for all intents and purposes, there really has not been any marketing approach. It is an industry where most of the business is done almost by accident. It has been, with very few exceptions, almost a cottage industry. A good friend of mine, a man who is a high-level marketing executive with a Fortune 500 company, said to me one day, after I had described for him the ineptness of my chosen business, that it was an industry suffering from "delusions of adequacy."

Now, for fifteen years I've made a *better* than good living in the security business. I own a home that's worth well into six figures, my wife drives a costly Mercedes automobile, my two sons attend expensive universities, and I have a safe-deposit box crammed full of certificates of deposit. I am 49 years of age, and if I so chose, I could retire right now and live comfortably for the rest of my life. At this point, you're probably saying to yourself, "If this guy has done so well in the security business, why is he knocking it, and, further, why is he taking the trouble to write this book?"

Perfectly legitimate questions. And I'd like to answer them frankly and honestly. From my answers, I believe you may be able to gain certain insights and motivations that will enable you to make a great deal of money in the security business, if you so choose, and if you're willing to pay the price. I wrote the book for a combination of reasons—some self-serving, others altruistic, that is, for the benefit of others.

Over the years, as I've traveled the length and breadth of the United States (and other countries as well), working at my profession—the training of security systems salespeople—I have been asked thousands of times, by newcomers to the field and by those not-so-new to the field, "Where can I find a book that will provide me with the techniques, methodology, and philosophy that will enable me to become a true professional in the science of selling—and, specifically, selling security systems?"

Yes, selling is a science (not an art), and yes, it is a profession. But all of the other sciences and professions—including selling—have connected to them large bodies of written works delving into every conceivable aspect of the particular science.

Not so, however, with that sub-branch of the science of selling concerned specifically with the selling of security systems. There are dozens of excellent books on selling, some truly brilliant, written by genuine masters. And many of these should be on the bookshelf of anyone who wants to become a legitimate professional in the field of selling.

Every true professional, be he doctor, lawyer, architect, etc., builds and maintains a library of books concerning his chosen profession. So should you build your own library of books on the science of selling if becoming a sales professional is your goal. But up to the point when this book was written, you couldn't add to your library a book on the science of selling security systems because such a book didn't exist.

Of course, I didn't write this book simply to fill spaces on bookshelves around the country. I wrote it because I saw the need. And I realized that if I were the one who successfully filled that need, I would make serious quantities of money. Money, in the business facet of our society, is the necessary gauge of how well someone manages to fill a need. And I like to make money. It's fun, it's satisfying, it's ego gratifying, and it gives me the options and freedom to do a great many of the things one simply can't do without it. And, frankly, while money is the prime motivator for this book, there are other motivations over and above it, and possibly of even greater importance.

Nothing, I believe, can match the gratification one can get from helping others to set goals and helping them to equip themselves to reach those goals. I could never adequately describe the satisfaction I receive when I meet a person who attended a seminar or training course that I gave and tells me how the concepts contained in the seminar, when implemented, dramatically changed his income and, in essence, his future. In fact, it is a glorious emotional "high" for me when I meet someone who, before I met him, had no future—or a mediocre one, at best—and after spending time in my training course or seminar suddenly had the insights and the knowledge to guarantee a future filled with constantly increasing income, job promotions, career satisfaction, and all the other things that make up the lives of the dynamic "go-ahead" people. It is enormously satisfying to be able to contribute to making a person a "go-ahead" instead of a "stay-behind."

It is truly a joy to help someone who has chosen to become a security systems salesperson and is "accidentally" selling *X* number of systems per week to become a sales professional who "on purpose" sells *X* + 3 or *X* + 5 security systems per week. It's an almost equally pleasurable accomplishment to help a veteran security systems salesperson to expand the outer limits (or what he thought were the outer limits) of his earning capabilities.

What pleasure to help some new entrepreneur who has entered the security business to create the basis for building a business that cannot help but grow and prosper. That basis is professional selling skills—skills possessed by (taught to) the entrepreneur himself and those people he subsequently hires to compound and expand upon his own efforts.

But even these satisfactions and pleasures are, in their way, selfish. While they don't specifically produce financial rewards, they are still personal rewards—something I receive.

Over and above personal rewards of this type, be they financial or psychic, there is a facet of the work we do (selling security systems) that is really a public service of the highest order. It is a public service that is rarely recognized but is, in reality, quite profound.

The value of many activities commonly referred to as public service is open to some question. But about the work we do—helping people make positive decisions about acquiring intrusion protection for themselves and their families—there can be no question. Every family motivated to protect itself is one less family that's a potential candidate for becoming a crime statistic.

The police and other government institutions have been powerless to stem the tidal wave of crime that has been engulfing this country. No one—no institution, no political body—has been able to do anything remotely effective in halting the flood of crime. In fact, throughout the nation law enforcement budgets have generally been cut. The police are literally drowning in crime.

The citizenry has done almost nothing to protect itself. Ninety-eight percent of the homes in this country have no security system of any kind. Less than 40% of the businesses in this country have taken any steps to protect themselves. People really don't think it will happen to them. They believe it always happens to "the other guy" and forget that to "the other guy," they are "the other guy."

Statistics indicate that 30% of the households in the United States will be victimized by a crime in the course of a year. And yet, almost no one voluntarily takes any steps to protect himself. It's the human condition to procrastinate; to plan to do something about protection "tomorrow," or "next week," or "soon." The successful security systems sales professional dedicates himself or herself to helping citizens realize that it can and probably will happen to them and, in so doing, helps them to make positive decisions about acquiring protection.

But since all humans resist making decisions (for fear of making the wrong ones), it requires the selling skills of a professional salesperson to motivate more people to make more positive decisions than negative ones; that is, to help a higher percentage of the prospects talked to decide to *acquire* a security system than to decide *not* to acquire a security system. The number of positive decisions obtained by the salesperson, in relation to the total number of times the salesperson "tells his story" to prospective buyers, is called "the closing ratio."

This entire book is directed toward creating consistently high closing ratios for security systems salespeople—in other words, how to get a "yes" instead of a "no."

And don't, for a moment, think this will happen "by accident," or because you "have the gift of the gab," or because you're "a born salesperson." It won't! It will only happen when you acquire, and are able to utilize, professional selling skills. And, as with all professional skills, you're not born with them—you must learn them. Once you learn them, you must practice, drill, and rehearse them. You must constantly hone and refine these skills so that, like a professional athlete, you are always in "top form."

And, like almost all learning in this world of ours, you'll eventually end up being graded—a report card, if you will—except in this case it's called a W-2 form!

So, welcome to my world. Welcome to the world of the professional; welcome to the world of impressive earnings, psychic gratification, and true public service.

2

Who Should Read This Book?

Chasing the Dream, and Catching It, Whoever You Are!

Let's talk about two broad categories of people who should read this book. And then let's talk about a number of more specific categories. The two broad categories of people I'm directing this material to are (1) those who are already in the business and (2) those who are contemplating entering the business.

If, on the one hand, you've been in the security business (for whatever length of time and in whatever capacity) and have the feeling that you should or could be doing better financially, this book is for you. It's for you because it will help you get considerably closer to the outer limits of your capabilities and your capacity to achieve. If, on the other hand, you don't have the feeling that you could or should be doing better, then you have an ego problem and should be reading another kind of book!

This book—and the techniques and methodologies described in it—is guaranteed to help you (or people who work for you) sell more security systems. I already know that, and you soon will, after you read this book and begin to implement the concepts contained in it.

There may seem to you to be a touch of arrogance in the preceding paragraph. But let me assure you that it is not arrogance at all, but confidence in the effectiveness of these classic and proven concepts. You see, I've seen them produce results in thousands of instances over many years.

Typical is the case of the fledgling saleswoman who worked for a Philadelphia alarm company and who, after less than a year in the business, was discouraged and ready to "throw in the towel" because she was averaging one to two sales per week. As a favor to the president of the company for which she worked, I agreed to spend a day with her, a day I spent illustrating professional selling techniques, outlining the methodologies for dealing with objections, and providing her with insights into buying motives, closing methods, and a host of other professional selling skills. I wished, at the time, that she and I could have spent more than a single day together, but the press of a busy schedule prevented my devoting more time to the task.

The point of this vignette is that in the third week after she shook my hand and thanked me for the time I had spent with her, she sold seven security systems and her commissions for that week were almost $2,000. She now averages sales of between five and seven systems a week, and her W-2 form this year will evidence the kind of income most people only dream of.

This woman, and the hundreds of other men and women whose cases are documented in my files, are not figments of my imagination. They are not fictions contrived by me to "toot my own horn" and to further my own self-serving ends. No, they exist, they are all real, they are really doing it!

And, the irony of it all is that none of these people was born with any special gifts, any special talents, or extraordinary intelligence. No, they were simply people who were money motivated and at some point realized that if they were ever to have the incomes of professionals, they must acquire the skills of professionals. They realized that the highly paid surgeon must first spend years in medical school and in internship and residency before he can afford to live in the big house on the hill, drive the expensive sports car, and send his children to good colleges.

In an oft-quoted remark, the Duchess of Windsor once said that "one can never be too thin or too rich." This, obviously, is not the forum for commenting on the desirability of thinness—or the lack of it—but it is most assuredly a forum for commenting on matters financial. Now, it goes without saying that *rich* is a relative term. What is rich beyond the imagination of one person might seem like petty cash to another. The point here is that even if you're not a fledgling or a novice in the alarm business—even if you're a long-time salesperson who has made a respectable living in the business; or maybe the owner of an alarm company, one who has "done all right" for himself—you must strive for growth. In business, if there is no growth there is stagnation, and stagnation is another word for mediocrity.

Mediocrity is a dangerous—though common—condition. Mediocrity is dangerous because it can't combat new competitors, new technology, economic recession, and the host of other "demons" constantly waiting around the corner from every business. As I said, mediocrity is stagnation. And stagnation leads to regression. And the old bromide we've heard so often is so true: "The business that doesn't go forward goes backward." Nothing is forever.

Even if you're a second or third generation owner of an alarm company who inherited the business from his father or grandfather, if you don't grow, new competition will eventually come onto the scene and begin to eat into your share of the market. This will happen regardless of what city you're in and regardless of how dominant a position in that area your company might have enjoyed for decades. The things that worked for your father won't "cut it" in the 1980s and 1990s.

If this particular shoe fits, and you're content to sit at home and clip coupons from tax-free municipal bonds, God bless you—have a nice life! I can and must then tell you that you've wasted the time and money you've invested in this book. So much, temporarily, for those who are already in the alarm business, whether salespeople or management. We will momentarily get back to this category and talk about some specific groups within the category.

Let's talk briefly about people who are not yet in the business. Possibly, you're now involved in selling. You've seen the crime problem engulf the country, and you've said to yourself, "Someone who was even half a salesperson could make a great deal of money helping to fill the obvious need of citizens for protection."

Or maybe you're a person who is not doing as well financially in selling as you think you should and are thinking, "If only I had the right product or service to sell, it would be a whole different ball game."

It may be that you make your living in some way other than selling and have always thought you could be good at it, provided you were to have a product/service to sell that could excite both you and your prospective customers.

Perhaps you own a business that doesn't satisfy you either financially or psychologically, and you're looking to enter a new business with more potential and satisfaction.

Maybe you own a business in partnership with someone else, and that someone else doesn't see the future as you do. If you are part of a partnership in which the partners are "out of sync," every day can be a genuine trial. Such a situation can be just as sad and debilitating as a marriage in which the man and wife are not in harmony.

Maybe you're a young person, recently finished with your formal education and searching for a "calling"—a profession in which you can not only make a good living but also satisfy your needs psychologically and intellectually. Who knows, you may fit into several of the foregoing categories, or maybe you fit into one I haven't even thought of yet.

But wherever you fit, whatever you are, or think you are, I have some caveats—some warnings—for you. I also have some basic principles to impart to you, principles that, if followed zealously, can help you to attain a level of professionalism that will, by its nature, enable you to earn the kind of income true professionals earn. Such professionalism can also bring you the satisfaction that derives from being among the best at what you do. I have these warnings for you because I know from years of experience that there is a world full of misconceptions about selling generally and certainly about the selling of security systems.

I continue to use the term *professional*, but I should really never use it alone, it should always be accompanied by *hard working*. It's not a professional who will make the "serious money"; rather, it's the "hard-working professional." You must understand, up front, that *selling is the world's best paying hard job and, conversely, that it's the world's worst paying easy job.*

Even the famous turn-of-the-century sales "great," Diamond Jim Brady, probably one of history's biggest earning salespersons, modestly attributed much of his success to luck saying, "The harder I work, the luckier I get."

Don't read this book if your conception of the working life of the salesperson is one made up of simply getting all dressed up in a stylish suit, eating leisurely lunches in nice restaurants, being charming, and regaling prospective customers with stories and anecdotes, and then having them volunteer to buy what you're selling. Obviously, I'm exaggerating, but you'd be surprised at how many people have a perception of what professional selling is, which comes very close to what I've just described.

Don't read this book if you think that all you have to do is learn all about the various types of security systems and how they function and simply tell prospects about them—again with the prospects volunteering to buy from you.

I can tell you, in advance, they won't. I can tell you about multitudes of people I've met throughout my years in the security business—people with prodigious knowledge of the workings of security systems, people who never made anything remotely resembling a professional-level income from selling systems. When most of these people did succeed in selling a system, it was more an accident than a purposeful, professional, targeted sales presentation.

I've spent many years shaking my head in sadness when contemplating the fate of thousands of men and women who have been hired by security systems firms, ostensibly to be professional salespeople, and who, because these firms themselves had not the vaguest notion of what professional selling skills are, let alone how to teach others to implement them, came out of the starting gate doomed to mediocrity at best, failure and discouragement at worst.

This lack of knowledge of professional selling skills, and the consequent inability to train new salespeople in these skills, has kept the vast majority of security systems firms small. They can't grow because they can't build a sales organization.

In thousands of instances, the start of a new alarm company results from an installer deciding that he wants to be in business for himself. This typical entrepreneur knows a great deal about alarm systems and is very good at installing them. So he enters the alarm business thinking that he'll be able to sell enough systems to sustain himself, make a living, and maybe even grow.

Even though this fledgling entrepreneur has not the vaguest notion of professional selling skills, he probably can sell enough systems each week to sustain himself. One system per week, if it's large and profitable enough, can put bread on the table. And, over the past decade, the alarm systems business has been so good that one could achieve this kind of sales volume almost by accident, if only one were willing to get out of bed each morning. But depending on accidents and luck to generate business is a dangerous and treacherous basis on which to attempt to build a business. These dangers are attested to by the high mortality rates among new alarm companies.

You can prove this to yourself by performing a simple experiment. Take the Yellow Pages or the business phone directory of any city in the United States. Then turn to the section titled "Burglar Alarms." Begin dialing at the AAA Burglar Alarm Company, and go through the entire category (and in any city of even moderate size, this will be quite a lengthy list) all the way to the end, the Zebra Alarm Company. You'll discover that fully 30% or more of the phones are no longer in service. Yes, that's correct, approximately 30% plus of the alarm companies listed in the previous year are no longer in business. I've performed this experiment in dozens of cities and have never been off by more than 5%. And scores of people to whom I've suggested the experiment attest to the validity of my statistics.

What do these frightening failure statistics tell us? They

tell us a number of things. They tell us that, while the alarm business is an "easy entry" business, like all easy entry businesses, it is also a high failure rate business.

A person can literally go into the alarm business with zero capital. Working out of one's cellar, garage, and/or automobile trunk, one can go out and sell an alarm system, take the customer's down payment, and go to a local wholesale alarm components distributor to purchase the various needed parts, install the system, collect the balance due, and then start all over again. This means that no overhead is required, no inventory is required—nothing is required except some elementary tools and some sweat and muscle.

Thousands of alarm businesses have been started in just this way. Many "made it" and managed to survive and enjoy some growth—thousands failed—and this cycle will continue ad infinitum. In the majority of cases, those that survive continue to consist of an owner or partners who both sell and then install the systems they sell. In another large group of companies, once some modicum of growth has occurred, it becomes an owner selling, with a hired installer(s) to put the systems in.

The next stage of growth, once some degree of longevity has been attained, and some of the desired growth, is for the owner to seek to hire a salesperson. This is a critical stage in the life of an alarm company. This is the transitional step that may determine whether the company will grow larger or remain a one-man band.

The owner, in the vast majority of instances, tends to hire in his own image. Since the owners themselves are not true professionals in terms of selling skills, they are generally at a loss as to how to train their first salesperson. In most cases, they simply try to teach the new salesperson all about alarm systems in a week, then throw them out into the street with the usual pat on the back and a "Go get 'em, Charlie!"

If the first salesperson to be hired is stolen or seduced away from another alarm company and is thus ostensibly "experienced," the chances are all too good that the company from which he came is equally deficient in any form of professional-grade sales training. I make this statement with little fear of contradiction, since the number of alarm companies with any kind of genuine professional sales training is minimal.

Owners of small-to-medium-sized alarm companies who are attempting the beginnings of a sales organization are probably some of the ultimate examples of "the blind leading the blind." Generally the results are considerably less than what could even be generously labeled successful. And that's the reason, basically, that so many alarm company owners I've encountered over the years tell me how little success they've had in attempting to start a sales force—and how discouraged they are at their difficulty in getting bigger.

One of the roots of the problem is that most people in the alarm business are product oriented rather than sales oriented. Think of it this way: Most companies in the business refer to themselves as alarm systems *installation* companies. This is their problem. The fact of the matter is that they should perceive themselves as alarm systems *sales* companies (who happen to have to install what they sell). If they thought of themselves in this way, they could not only *be* a different kind of company, they also would *attract* a different kind of salesperson.

What the mainstream of the alarm business has traditionally failed to bring into proper perspective is the fact that money isn't made *installing* alarm systems; it's made by *selling* alarm systems. Even the successful alarm companies have failed to understand this philosophy. *Successful*, of course, is a word that means different things to different people. In terms of numbers of systems sold, you could count on the fingers of a barbershop quartet the number of independent alarm companies that sells 25 systems a week. And this includes companies that have been in business for generations.

And, to boot, the bulk of the business that even the most "successful" alarm companies do is primarily commercial and industrial business. The alarm industry has been almost totally unable to "crack" the residential alarm market. The percentage of market penetration in the home market is so low as to be almost an embarrassment. I know of companies, divisions of major corporations, that still, in spite of their huge size, lack the true insights to seriously penetrate the residential alarm market. The proof of this is the abysmally low market penetration rates.

The point of all of this is to say to you, whoever you are—whether you're in the business, planning to enter the business, or even planning to leave the business—don't think you can make serious money or build a meaningful organization in the alarm business without the utilization of professional selling skills and real insights into what it really takes to motivate *large* numbers of prospects to buy alarm systems.

If you're the owner of an independent alarm company, ask yourself why your firm *doesn't* sell 25 systems a week or more. Ask yourself why you don't have any salespeople who sell a system (or more) *every* day.

If you're an alarm company sales manager, ask yourself the same questions.

If you're the head of an alarm division of a major corporation, ask yourself the same questions. How many branch offices do you have that sell 25 systems a week? Do you have any salespeople in your organization who sell 7 to 10 systems a week?

If you're the head of the largest alarm systems sales company in the world, ask yourself why your company has had such an *underwhelming* track record in penetrating the residential alarm market.

If you're a fledgling alarm systems salesperson, ask yourself if the company you're working for has the know-how to train you to become a real professional. If the answer is no, find another company to work for. If you don't, you're doomed to remaining a member of the medium-income category, and that's not much of a future to look forward to. Better yet, spend a year implementing the techniques outlined in this book, and you'll end up "owning a piece of the joint." They won't want to lose you!

Please don't misinterpret all of this as arrogance. It's not. It is the confidence that comes from knowing that what you're saying and what you're telling people to do *will work* and will help them to become the successful professionals all of us have the power to become.

I truly wish I could transport each and every one of you

to the headquarters of an alarm sales company in one of the boroughs of New York City—probably one of the toughest and most competitive markets in the world—a marketplace with hundreds upon hundreds of alarm companies, some of which have been established for generations. The man who owns this particular alarm company founded it in 1982. He was a complete novice to the alarm world. What he had was the drive and the motivation to attempt to create a successful money-making machine based upon his realization that the market potential was gigantic, and that what was need was a genuinely professional sales organization executing a professional sales plan daily. And now, barely 24 months later, this man runs an alarm sales operation that rarely sells fewer than 30 systems each and every week. He employs eight salespeople. Yes, that's right—30 systems each *week*. And they're excellent systems sold with a generous gross profit margin.

My point in telling you this is my pride in knowing that the basis for this man's success is the fact that his sales effort is based entirely on the concepts contained in this book. His success, and scores of similar stories of companies and individuals who have used the techniques and insights contained here, are all documentable. They're not fiction!

And so I say, when asked "Who should read this book?", anyone—whether an owner, a sales manager, a fledgling salesperson, or an experienced salesperson—should read this book. Learn the techniques. Absorb the insights. Capture the philosophy. And then internalize it all! By that I mean make it all a *part* of you; make it all reflexive. Then go out and practice. Drill and rehearse it.

There are very few "sure things" in this world, but one thing I will guarantee you, if you do as I suggest, is you will sell security systems *in quantity*.

If you're an owner or sales manager, this book can become the text and curriculum for your sales training program. It is currently being used by hundreds of alarm sales companies in the United States and a number of foreign countries. The techniques are all here. The actual words to say are here. The only other ingredient required is *you*. Use these techniques—say these words—and I can promise you unequivocally that you can build a business, make impressive amounts of money, and dramatically change your future.

3

Why This Is a Great Business To Be In

The Market and the Market Potential for Security Systems

In the last quarter of the twentieth century, one would think it almost impossible still to find a business that combines an almost untapped market with a market climate as near to perfect as can be imagined. A person would be hard pressed to find a business with more potential for serious money making than the security systems business.

If such is the case, you may ask, then why the frightening mortality rate of security systems companies, which we talked about earlier, and why the general lack of successful entry into the business by major Fortune 500–type corporations? These are legitimate questions, and the answers to them simply serve to reinforce the excitement about the potential. But this potential exists only for those who approach the business equipped with the proper insights, the proper philosophies, and the intelligence to disregard the security industry's conventional wisdom.

The intelligence to disregard the conventional wisdom is critical because, remember, the foolishness and lack of validity of this "wisdom" are exactly the reasons why the wide-open opportunities are present in the security industry. If the conventional security business wisdom were valid, the market would be penetrated to a significantly higher degree—possibly even near saturated.

Whether you're a salesperson, a sales manager, an owner or an executive of a corporation who is contemplating entering this business, I guarantee you that if you pay heed to the old bromides, the clichés, the fish wives' tales so prevalent and so long held as gospel, the visions of sugar plums you might have will, in short order, turn into moldy figs. But before we begin, in earnest, to make our case for the assassination of the many and varied sacred cows in the security business, we should spend some time discussing the size of the market and the *real* market potential.

There are 81 million residential dwelling units in this country. Less than 2% of these units have any kind of security system. This means that there are, technically, somewhere around 79 million potential prospects for home/residential protection of some kind. Over and above this number are the over 1 million housing starts each year. Thus, over a million new prospects are coming on-stream each year in the residential sector of the market alone.

Add to this over 14 million businesses in the United States, with over 6 million of them having a net worth of over $25,000—of which over 75% have *no* security system. No one really knows how many new businesses are started every year, but we do know that there are over 500,000 new incorporations each year. Obviously, hundreds of thousands of additional businesses are started that do not incorporate themselves. Whatever the totals of these categories, suffice it to say that the number of potential security systems prospects is so huge that it literally takes one's breath away.

Certainly, these big numbers can only be considered market potential in a technical sense. Obviously contained in these huge figures are vast numbers of prospects who might want to buy protection in the worst way but simply can't afford it, believe they can't afford it, haven't been offered something they *can* afford, or haven't been motivated to think they *need* it.

Talk to any number of people in the alarm industry, and you'll get any number of opinions about how big the market potential really is and just how much of that market has actually been penetrated. But think about it for a moment. Does it really matter whether the market potential is 81 million or 73 million homes that are unprotected? Does it really matter whether there are 8 million or 6 million businesses unprotected? Of course it doesn't. The practical fact of the matter is that whatever the true numbers are (and no one really knows), there is such a small degree of market penetration that it would take decades to reach anything remotely resembling saturation.

Further, the market keeps getting bigger. There are, even in an economic slump, over 1 million new housing starts each year. And as we noted earlier, hundreds of thousands of new businesses are initiated. So what we have is a situation where the market potential increases by the millions each year, while the market penetration (alarm systems sold) increases only in the hundreds of thousands. The two figures will *never* meet. The market will probably forever remain relatively untapped.

Compare this with big-ticket home appliances, for instance, where the percentages of market penetration are often in the 80% and 90% areas. Over and above big-ticket appliances sold into the new housing market, the rest of the market is a replacement market. A thinking person wouldn't look to the big-ticket appliance market for new opportunities, would he? He wouldn't, if the term *thinking* accurately described him.

Obviously, the biggest opportunities lie in those markets that are relatively unpenetrated, although being untapped is only one of the things we should look for in seeking big money-making opportunities in markets. Being untapped is an academic virtue if the populace doesn't need and/or doesn't want the product(s). We'll talk here, momentarily, about the market as it relates to people's needs for alarm systems, and we'll talk in a later chapter about buying motivations and the public's wants for alarm systems.

Another profound virtue in the burglar alarm market is the competitive situation. As opposed to most other industries, where the competition is sophisticated and comprised of corporate giants that in many cases sell their wares through thousands of retail outlets, the alarm industry is almost a cottage industry. There is no retail competition in the alarm industry. Every attempt to market home alarm systems via conventional retail channels has been little short of a fiasco. It would be foolhardy for a person to attempt to enter the personal computer business today and attempt to compete with the multi-billion-dollar giants like IBM or Apple, or any of the dozens of other large makers. One also would probably be equally foolhardy to attempt to enter the personal computer retailing area, where chains of hundreds of stores saturate the countryside.

No need to mention the jungle-like competition in the appliance business, with thousands of retailers dotting the landscape, selling their merchandise at almost microscopic profit margins. This is another aspect of the competitive situation in the alarm business that makes it so attractive. Because the competitive ambiance is so primitive, there are lush profit margins to be had for the astute sellers. Whether you're an owner or an alarm systems salesperson, the profits should be excellent if you generate the sales volume and charge properly for your systems.

If you're an alarm systems salesperson and are a volume producer, you can command much better than average commission rates. The key phrase here is "volume producer." If you produce volume, you can write your own ticket. Alarm company management needs you, and if you're good they will pay extra to keep you. If they won't, go elsewhere!

Don't misinterpret that statement as arrogance. It's not; it is simply an expression of the reality of the matter. If you're good—super good—you can go anywhere and be welcomed as the producer/professional you are. But, of course, you won't be super good unless you make yourself a true professional.

It should be noted here that selling, per se, is probably *the* most potentially rewarding (financially and otherwise) field of endeavor in relation to capital investment. The capital investment, in fact, is zero and the ceiling on your income is unlimited.

If you're selling (or thinking of selling) for an alarm sales company, don't think that your income ceiling is the income of that company's top producer. Who says it's not possible to earn more? To earn more in the alarm sales profession, you've got to develop more competence. That's the simple secret. And that comes only from learning, studying, practicing selling skills, drilling, and having product knowledge, competitive knowledge, and people knowledge—all the ingredients that, when combined, spell competence. Make no mistake about it, if you are competent you will stand out as an island in a sea of incompetence! And you will make money—serious money—because of your competence.

Your competition in the alarm business, for the most part, is made up of people of little or no sales competence. It's been an old-fashioned industry that, like many others, did not understand how to train professional salespeople, didn't really believe in it. It's been an industry that concerned itself with hardware and technology and that ignored professional selling skills. This is because the vast majority of the people who make up the industry are not professional salespeople. They have shied away from selling skills training because they were ignorant of it and so took refuge in the hardware.

For many of the earlier entrants in the game, business came as a result of people who approached *them* requesting to buy an alarm system. And there were always some referrals. This was enough to make a fairly good living. Added to this were the businesses that were forced to buy alarm systems by their insurance carriers. In essence, it's an industry that didn't really sell but that made a living by "accommodating volunteers."

One of the primary reasons for the higher degree of alarm penetration in the commercial segment of the market is simply that there were more voluntary buyers from that segment. The older, larger alarm sales companies have been built almost exclusively in the area of commercial alarm systems.

For years, the industry talked about the obviously larger and obviously less penetrated residential alarm market, but this segment has remained virtually untapped. This is because the industry didn't understand how to get into the homes of America and knew even less about what to do if they were fortunate enough to get into a few. *Fortunate* is the key word here because that's generally what it has been: fortune, or luck!

Not planned implementation of professional selling skills and expertise.

The point of all this is to say to you that if you'll develop your professional selling skills—no matter where you're operating in the alarm sales business—you'll be a member of a select elite and can get paid off accordingly. Go the extra mile and you can be a superstar and be rewarded at superstar rates.

Most of your competition doesn't believe in what this book is all about. That's why they've been "so successful" in obtaining such "tremendously high" penetration of the alarm market. We entitled this chapter "Why This Is a Great Business to Be In." Well, among the reasons is that your competition are primarily amateurs—and amateurs trained by amateurs: the blind leading the blind.

I couldn't begin to count the instances, in my years in the alarm business, when I was invited to observe so-called training sessions. And what I observed was people teaching other people *how alarm systems function* and how they're installed. No mention of how to motivate prospects to make positive decisions about getting themselves protected! No mention of how to answer all the inevitable objections! No mention of how to close! No mention about how to close . . . now . . . today . . . tonight! No mention of the one-call close!

What we're really saying here (at the risk of belaboring the point but important enough to risk it) is that even if you're only *part* of the way along the road to becoming an accomplished professional alarm salesperson, you can outshine any competition you encounter. As some unknown philosopher once said, "In the kingdom of the blind, the one-eyed man is king."

Your professional competence obviously will be the ultimate producer of big money results, but in the alarm business you've got something else going for you—something that is unique to our industry: you've got a universal *need* resulting from a monstrous frightening crime problem. The scourge of crime is truly democratic; it strikes anyone, anywhere. Crime strikes regardless of the victim's race, creed, or color, regardless of time, geography, or economic station.

You've got a market with a universal need but not a universal *want. Everybody* needs an alarm system. Yes, they really need one, but relatively few have been motivated to make a positive decision to acquire one. What a marvelous thing for a salesperson to contemplate: "Everybody needs what I'm selling." And all you've got to do, as a competent professional, is understand what it takes to motivate prospects to *want* what we already know they *need.*

This is quite different from the multitude of products that consumers buy by the millions—none of which they really need—products like videotape recorders, microwave ovens, personal computers, video games, etc. Nobody needs any of these products and dozens more like them. Consumers simply *want* them!

You may ask, at this point, why one should go through all the necessary gyrations involved in attempting to persuade people to want a security system, when it might seem to make more sense (and be easier), to sell them some product they already want. The answer is that while it might, just might, be somewhat easier to sell people things they already want, it's not nearly so well paid an effort.

How much could you be paid as a sales clerk in a personal computer store? Not very much, you'll agree. What are your profit margins as a retailer of videotape recorders? In the retail jungle, some retailers are discounting prices to the point at which they're working at 5% to 10% over their cost. And the retailer must invest substantial amounts of money in incentives, fancy stores, overhead, etc. Further, wants are changing constantly; fads, vogues, fashions come and go.

But the legitimate, genuine need for security systems never changes; in fact, it actually becomes more critical as the months and years go by. Why? Because the crime problem not only is ever present but it gets worse. Everywhere! Burglary, particularly, is pervasive, and no one really knows how many burglaries actually take place in the United States each year. Neither local and state law enforcement agencies nor the FBI really know the extent of the burglary problem.

Local and state police tally up the number of *reported* burglaries each year in their bailiwicks and send these statistics to the FBI that, in turn, publishes what amounts to an annual report on crime in the United States. This, of course, is a report totally without any validity since it provides data only on crimes that are reported, not crimes committed.

A U.S. Department of Justice study, begun in the mid-1970s, has determined—through interviews with tens of thousands of Americans—that only about one out of every three crimes is reported. This means that if, for instance, the FBI crime statistics show somewhere over 3 million burglaries committed last year, the likelihood is that the true figure is somewhere over 9 million burglaries—three times the number reported.

An ironic footnote to this subject is the first page of last year's FBI *Uniform Crime Reports*, in which the bureau's director said, in essence, "We don't know what causes crime, and we really don't know what to do about it." While the statement may be ironic, it certainly is not a news flash. From the beginning of recorded time, there has been crime, and no one ever had a solution, nor *will* anyone ever!

Over the years, as our country made the transition from being a rural/agricultural society to an urban/suburban society, and the total population grew enormously, crime grew proportionately, maybe disproportionately. No need even to mention the big city crime problem. It is on the front pages of our daily newspapers every day. And the suburban and rural crime problems are equally frightening.

As the suburbs grew from being merely bedroom communities into self-contained mini-cities with their own business and industry, they also began to grow their own criminal elements. Thus, the citizens who fled the dirt and crime of the cities ultimately discovered that there was truly no place to hide. The burglars and stick-up men were no longer coming out of the cities to prey on the suburbs; they were already there.

And so, today, the druggist who wouldn't dream of leaving his store in the city unprotected is equally concerned about protecting his residence in the suburbs. The suburbanites are frightened!

And this fear, this paranoia, has created what must be as close to an ideal market climate for burglar alarms as could be possible. And yet, 98% of the homeowners in the United States have no security system! The reason? While they may be nervous and frightened, most people manifest the all-too-common human failing: procrastination. Combined with this is the fact that, while most people can intellectually accept the statistical reality of the danger of burglary and/or robbery, they feel that it is the other guy who makes up the statistics. Few people can manage to envision themselves as statistics. They can really only be moved to making positive decisions about getting themselves protected when they are motivated by *fear.*

We'll delve deeper into buying motives later; however, the important thing to point out here is that the primary reason for the almost ludicrous lack of market penetration is that those in the alarm business, with a few exceptions, never really understood what it takes to move large quantities of people to make the decision to get themselves protected. What this means to you is *opportunity.* If you personally understand what it takes, and if you're willing to develop your selling skills to a professional level, you can "eat your competition's lunch." In fact, as a practical matter, for you there will be no competition.

While you are busy prospecting, skillfully communicating the concept of "peace of mind" and motivating prospects to say yes, your competition will be waiting for *volunteers* to whom they will attempt to sell some security hardware. It's no contest!

In the vast majority of instances, you will literally have no competition because you will have generated the prospect yourself and will actually end up being on stage in front of the prospect solo. That's the beauty of direct-to-the-user, eyeball-to-eyeball selling: you're on stage *solo.* If you're a pro and succeed in closing in at least one out of three interviews/presentations, you'll be instantly rewarded. When you fail to close, you're penalized immediately. The bottom line becomes evident on the spot. Isn't that great?

You're either already in, or contemplating entering, a business with more built-in success potential than probably any other business in the world today because:

The market potential is huge, yet almost untapped.

The crime problem is not only epidemic in proportion but also universal in geography. The market is receptive, frightened, and it's everywhere.

The competition is highly fragmented, unaggressive, and considerably less than professional in terms of selling skills and insight into buyer motivations.

What are you waiting for? It's time to get rich!

4

Who Are You and What Do You Want?

Getting Rich Selling Security Systems by Solving Your Identity Crisis

We must first ask you a question. If you cannot honestly answer this question with a yes, then I would suggest that you waste no further time reading this book, and immediately begin looking for a new vocation. The question is Do you sincerely want to get rich? Think hard before you answer, because it's not nearly so simple or so cut and dried as it may seem.

Most of the thousands of people whom I've asked this question over the years have almost reflexively answered yes. And yet most of them weren't even two steps down the road to the "big money." Most of them were producing substantially below their true potential. Why?

There are three primary reasons (and a multiplicity of secondary reasons) for the surplus of mediocrity (averageness) in security systems sellling. First, most people (both entering the business and already in it) don't understand or realize that selling is a science (not an art) and that the practice of a science is called a *profession.*

Before elaborating on this point, it might be worth defining these words. Following are Webster's Dictionary definitions:

> Science: Knowledge attained through study or practice; technique which may be studied and learned; like systematized knowledge.
>
> Profession: A vocation requiring specialized knowledge and intensive academic preparation; an occupation requiring training and advanced study in a specialized field.

Thus, a professional is one who has acquired the specialized knowledge and one *who has undergone the intensive academic preparation of his or her chosen vocation.*

What we're saying is that no one is born a professional. There are no born salespeople just as there are no born lawyers, doctors, architects, or any other brand of professional. These professionals are *made* through academic training, study, and practice. Internships and residencies are a part of all professions.

This book was not conceived in an ivory tower or a classroom. It was created in the field, using methods and techniques that have been *proven* to be effective and productive. There is no theory here—only fact and reality. The fact and reality for you, the individual, is to decide that:

> you really do want to get rich;
>
> you are willing to learn the concepts and techniques contained in this book;
>
> you are willing to practice, drill, and rehearse the techniques until you are ready to go into the field and implement them every day, every night, and every weekend and "make the numbers happen."

Make these decisions and you're on your way to becoming a true professional security systems salesperson and gain the kind of earnings such a person generates.

A second reason for so much of the mediocrity to be found in the security systems sales business is the lack of *disciplined* commitment to the philosophies necessary for major long-term success. Webster's Dictionary defines "philosophy" as "the investigation of causes and laws underlying reality; inquiry into the nature of things based on logical reasoning; any system of motivating concepts or principles."

And yet the myth of the "born salesperson" persists. Actually, it's a common excuse to avoid the hard work of learning. I've often heard people say, "You've either got it or you haven't got it." What they're really saying (whether they realize it or not) is "You're either willing to pay the price in hard work and learning or you're not willing." Everything—yes, everything—about selling is learned. Nothing about selling is instinctive!

So, the first decision you must make, if you genuinely desire the material rewards that professional security systems selling can bring you, is whether or not you want those rewards strongly enough to undergo the learning process necessary to become a professional. Don't kid yourself into thinking you can build a career at the professional level in selling by "winging it." If you allow yourself this delusion, you're doomed to spend the remainder of your selling career marooned on an island of mediocrity. This is the crucial aspect of your identity that you must "nail down." Do you want it badly enough to do what you've got to do to get it?

As I mentioned earlier, there are a number of near-masterpiece books on selling generally, but until now, it has been almost impossible to acquire the required specialized knowledge and to undergo the academic training needed to become a *true* professional in the field of selling security systems. It has been impossible because, unlike the medical profession, the legal profession, and all other professions, no real body of work (printed or otherwise) has existed pertaining specifically to the science of security systems selling.

This book, targeted toward the training of professional security systems salespeople, is among the *first* components of just such a body of works. These first endeavors are profound steps toward the future of the alarm industry because they are the beginning of the long overdue professionalization of the up-to-now catch-as-catch-can nature of security systems selling.

And what this really means is that new (and even veteran) home alarm salespeople can be trained to become predictably productive and produce alarm sales in quantity and on purpose, rather than once in a while, by accident. It means better salespeople with less personnel turnover because people who make serious money stay longer. It means the ability to build an alarm sales force via structured, formalized training. It means more sales and profits for alarm company owners.

In order to be seriously successful as a security systems salesperson or manager, you must be *philosophically committed* to:

The first-call close;

The critical need for a structured presentation;

Conceptual selling versus hardware peddling;

The realization that prospects buy alarm systems emotionally, not intellectually;

The need for a dramatic, nontechnical demonstration;

Multilevel pricing;

The use of the Security Survey;

The realization that there is only one motivator that moves prospects to buy now: fear!

We'll talk about each component of this philosophy in succeeding chapters of this book.

Without such philosophical commitment, my program (or any other) is an exercise in futility. *With* such commitment, you and/or your company can generate serious quantities of alarm systems sales on purpose and not simply a few systems by accident. There is no magic! There is no snake-oil panacea! There is only commitment to *proven* concepts and methodologies and the training of people to execute them in a professional manner. A training program (based on this book) is one primary tool for the implementation of your commitment.

A third reason we find so much mediocrity in the security systems sales business (all sales, really, not just security systems sales) is the inability of most salespeople to develop an effective system for dealing with *rejection*. One thing in sales is a given: You can't, don't, and never will sell 'em all.

No matter what your degree of proficiency as a salesperson, you ultimately develop a *closing ratio;* that is, you sell one person out of every so many prospects to whom you make a presentation. This becomes your closing ratio. Let's say, hypothetically, that you've developed a one-out-of-four closing ratio. You generally sell one security system out of every four selling interviews that you have.

This means a couple of things. First, it means that if you're maintaining this ratio with any degree of consistency, you're well on your way to becoming a pro, who is someone who can maintain (over the course of a selling career) a one-out-of-three closing ratio or better.

But what it also means is that for every one person who says yes to you, three people will say no to you. They will reject your efforts to help them make a positive decision about getting themselves protected. Notice I didn't say these three people (out of the four presentations) rejected *you.* No, they rejected your attempts to help them (possibly save their lives) for any one of thousands of possible reasons—all contained within them, *not you.*

If you're like most average salespeople, you find it difficult to handle the large doses of rejection that come along with a one-out-of-four closing ratio. Remember, such a ratio means that three out of every four prospects say no. If you're a one-out-of-four closer, you must make twelve presentations in a week in order to close three sales in a week. The key to the three sales is obviously the twelve presentations. It also means nine nos along with the three yeses. If you don't have a proper, effective way of dealing with the nos, you'll hide from them; you'll avoid making the number of presentations you must make to generate the three yeses. And, thus, you won't get them.

If you're to become a superstar salesperson, you must develop your own personal system for rejecting the negative effects of rejection. If you adopt my suggestions on how to do this, you'll not only be able to deal with rejection, you'll *learn to love no.*

Here's why no is good: Since you (hypothetically) have to make four presentations to make one sale, and three out of the four say no, every no brings you that much closer to a yes. If, on your second no, you realize that there's only one more no between you and a yes, you'll not fear another no—you'll be rubbing your hands in anticipation of it—because you know that after the next no comes the beautiful yes.

If you deal with no in this way, you'll have taken all of the pain out of the rejections. You won't hide out at home, in a coffee shop, running errands—doing anything to avoid making presentations (to avoid the pain of rejection). On the contrary, if you realize that every no brings you that much closer to a yes, you'll look forward to every presentation—you'll do sixteen instead of twelve—and make four sales instead of three.

If you once understand this concept, every time you're rejected you'll feel good instead of feeling bad. And you're on your way to making a fortune. Let's say it again: Every no brings you that much closer to a yes.

Here's another aspect of my formula for handling rejection that is equally effective in dealing with nos. Not only are nos stepping stones to yeses, they're also exceedingly valuable and you can put a literal dollar value on them. You certainly don't fear something that's worth money in your pocket, right? Of course not!

Here's what I mean and here's how it works. Let's say that your commission on the average sale is $200. Since you're a one-out-of-four closer, you've got to make four presentations to make one sale. Thus:

1 sale	=	$200
4 presentations	=	1 sale
Therefore, each presentation	=	$50.00

You're not paid by the sale, you're paid by the presentation. Each presentation is worth $50 to you. The nos are worth $50 each to you. Why would you fear a rejection that's worth $50?

The answer is you wouldn't. And once you adopt this concept, you'll find yourself making more presentations than you ever dreamed you could/would. Why not? They're worth money and you've conquered the pain of the rejection. Just think of it: every time you hear a no, you've earned $50.

Concentrate on the cash value of rejection, and you'll actually start looking forward to being rejected. Rejection is very painful to those who don't have a formula for dealing with it (as you do, now that I've given you mine), and what is even more damaging to a salesperson than the pain of actual rejection is the *fear* of rejection. If you're terrified of the *possibility* of rejection, you'll hide from the presentations (where the rejection comes from), and thus, by not making presentations, you'll not get the nos that bring you directly to the yeses. You critically need nos in order to get yeses.

Learn to love no. It's the key to the big money! Now, let's get on about the business of specifically detailing the concepts that, if you learn them and believe in them totally and practice them, will take you a long way down the road toward becoming a professional—with all of the rewards professionalism brings.

The CLOSING
CRIME
FEAR
COMPLIMENTS
RAPPORT
CREDIBIL
CONTROL

5

The Organized Sales Presentation

Touching All the Bases, Every Time

THE REASONS FOR, AND IMPORTANCE OF, AN ORGANIZED SALES PRESENTATION

Before getting deeply into the structure and the specific details of the organized sales presentation, it is crucial that we understand the *reasons* for its existence and the resulting *importance* of such a presentation. Once you understand these reasons, you'll realize that the organized sales presentation is the *foundation* upon which your effectiveness will be built *and maintained.* The presentation is *the* essential ingredient of a professional calibre closing ratio—that is, the number of *sales* made in proportion to the number of *presentations* made.

Let's first define what professional selling *is*, and from that definition we will see just exactly *why* the organized sales presentation is so very vital to determining whether or not you reach the earnings levels to which you should be aspiring.

"Selling is the presentation of a series of ideas in such a way that they lead people to action."

Well, very simply, once we have determined what ideas we must present to our prospects—ideas that will help them to make positive decisions about investing in security—and once we have determined in what sequence or order these ideas should be presented, then we organize those ideas into the sequence that, in essence, becomes our organized sales presentation. This organized sales presentation becomes the track we follow on each and every sales call. It is a schematic that guides us toward our ultimate objective: getting a yes from the prospects. Once we've established the structure and content of our organized sales presentation, we have the strongest tool in selling. We cannot be diverted from the path we must travel since we have a track to follow.

Those amateur salespeople who think they know enough to go into a prospect's home and "wing it" or ad lib their sales attempt, are doomed to a sales career (probably short lived) of mediocre performance and the accompanying mediocre income.

When we fail to use an organized sales presentation, we are almost guaranteed that our presentation will go off on to numerous irrelevant tangents that not only will divert us from our path—traveling toward a close—but also will bring about other negatives:

- Our presentation will end up taking too long. If we get diverted from our path, we can find presentations running to two and three hours. Since we have limited selling time in the evening, such a time frame would limit critically the number of presentations we can make. The structured presentation gets us back on the track quickly and saves precious time. It also gives us a *target* to get back to when we find ourselves diverted somewhere out into "left field." Otherwise, many times we will find ourselves groping to get back onto the path.
- We will forget to include key factors in the presentation. Because we're human, no matter how well we know our product and our business, we will (if we attempt to ad lib) constantly be leaving out key elements needed to bring

the prospect to a positive decision. With an organized, structured presentation, we are "locked in"; that is, as long as we follow the set structure, we will touch *every base, every time.* As long as we follow the preset sequence of our presentation, we are *taking the guesswork out of selling.*

We run the risk of losing control of the interview. Without a structured path, there is the danger that the prospect(s) will take control of the presentation; then, all is lost! We cannot allow them to take themselves where *they* want to go—we must lead them to where *we* want them to go—which is (through a preset sequence of powerful ideas) to a decision. By utilizing an organized sales presentation, we know in advance the script we are playing by, and thus, we have a strong advantage over the prospects and the tool with which to maintain control.

We have repeatedly mentioned the organized, structured sales presentation, and we've defined it as "a series of powerful ideas presented in a sequence so as to move people to action." At this point, you're no doubt thinking, "OK, what *are* those ideas, and what *is* the proper sequence?"

Momentarily, we're going to make it easy for you because we are going to supply you with the outline for a *tested and proven* organized sales presentation. Further, we'll give you the reasons why each and every one of the components of this presentation is necessary and important. Why it is there!

But before getting to the actual outline of the presentation, it would be well to comment about the manner of your delivery. Your speech and attitude must be professional but *conversational* and still always following a carefully structured presentation, or track.

Once you've been using this organized sales presentation for a short time, it will become internalized; that is, it will become a part of *you* and you a part of *it.* And, while structured, it will be conversational and seemingly ad libbed.

Certainly, at the front end, when you just begin to use this presentation, you will occasionally stumble. You will feel confined, but as you use it again and again, you will begin to feel more comfortable, and you will get smoother and smoother as you polish and refine.

You'll modify slightly. You'll inject your own personal style into the entire presentation, and you'll end up being just as conversational and natural as if you were in your own living room. You'll also become very excited when you see what the use of this presentation does for your closing ratio and, thus, your income.

From time to time, you'll hear someone refer to a "canned pitch." They'll usually say it with a sneer in their voice. You'll generally hear this from someone who is not a professional salesperson and someone who doesn't have even an inkling of what constitutes an organized sales presentation—what such a presentation is meant to accomplish and what it can do for a salesperson's closing ratio. If the person who uses the term *canned pitch* is in the alarm business, you can bet your bottom dollar that he or she doesn't produce quantity alarm system sales. What we're talking about here is an organized sales presentation that has been *proven effective.* A canned pitch is a speech, and it will *never* be effective.

Let's get on now with the organized sales presentation itself. First, we'll give you the outline of the presentation—that is, what the components of it are and what the right sequence is. Then we'll go back through the outline—point by point, segment by segment—explaining and giving reasons why each ingredient is included.

THE ORGANIZED SALES PRESENTATION

A. The Approach
   1. Organization
   2. Establish rapport
   3. Establish control
   4. Compliments
   5. Credibility
   6. Preempt certain common objections

B. Questions
   1. Establish fear index
   2. Establish alarm knowledge

3. Establish price perceptions
4. Establish budget perceptions
5. Establish buying motives

C. Establishing General and Specific Problems
1. Crime discussion
2. Localized discussion/examples

D. Offering a Solution
1. Demonstration
2. Security survey

E. Closing

Now, let's go through this outline, point by point.

The Approach

While all the components of the organized sales presentation are critical to its effectiveness, the approach may be the key to whether the interview will be a success (obtaining a yes) or a failure (a prospect saying no—or worse, saying maybe or "I want to think about it," etc.). Following are the things we accomplish in the approach segment of our presentation.

Organization. Prior to your arrival at the prospect's residence, you must do an *organization check.* You must make certain that your demonstration unit is in perfect working order. It should be spotless, immaculate.

Your paperwork should be crisp and orderly. Your survey and questionnaire should be right on top of the paperwork. Your leatherette/vinyl folder should have a legal pad in it.

You should look like the professional you want to be perceived as being. Remember, the prospects don't know you yet, so to them you are what you seem. Make certain you're dressed the way a professional dresses.

Establish Rapport. Webster's Dictionary defines "rapport" as "mutual trust" and "emotional affinity." Stated more simply, at the front end of our presentation we must establish good "vibrations." We must make the prospects *like us* and *trust us.*

We must establish a friendly, nonthreatening atmosphere so that the prospects will be receptive to what we're saying. We all remember at least one teacher in our youth who was friendlier and more likeable, and that teacher was the one we listened to and learned more from.

From the very beginning of the presentation, we must make quick, good friends with our prospects. We're not hucksters or pitchmen attempting to sell them something; we're friends of the family offering them good advice. To the degree that you are able to create an atmosphere of *rapport*, you will be able to gather better information and more honest, straightforward answers when you start asking your questions in the second segment of the presentation.

Establish Control. *Every* time you enter a prospect's home, a sale will be made. Either you will sell your prospect on the wisdom of making a positive decision to invest in peace of mind and security for his home and loved ones, or your prospect will sell you on why he should not make a yes decision.

The key element in shaping this decision is control. You must control the interview so that you can lead the prospects down the track of your organized sales presentation. Your control, in many cases, really takes the form of gracious questions; that is, "Would it be OK with you folks if we talked in the kitchen? I think we'll be more comfortable there, don't you agree?"

In such an instance, your control would be determining just where the interview would be conducted. By asking other questions, *you* end up directing where the presentation is going. *You* determine the direction and the sequence. *You* determine when the *transition* will be made from information gathering and so on to discussing the crime problems in the prospect's neighborhood, to a survey, to an attempted close. (We'll get to each of these specific segments shortly.)

It's really not difficult to control. We're low key and yet really in control. We're controlling with questions and suggestions. We know where we're headed—the prospects don't—and that is our advantage!

Should you lose control, there's no telling where the interview will be headed, but chances are, it won't be in the direction you want it to go. It then becomes a free-for-all, and as professionals, we can't allow this to happen.

The control factor—as with the entire organized sales presentation—is targeted toward taking the guesswork out of selling. Lose control and you lose your "edge."

Compliments. Everyone wants to be told that the things they own are attractive, in good taste, unique, or unusual. People will like us better (and more quickly) when we compliment them on things they own—for example, the manner in which their home is furnished or things like trophies, awards, paintings, photos, etc.

Make certain that you inject at least one or more complimentary comments during your approach. But make certain it's done in a sincere manner. There is no bigger "turn off" than the phony or insincere compliment. Such insincerity immediately stamps you as an uncaring person, not a real friend from whom to take advice.

Credibility. Establishing your personal and your company's credibility is such an important part of the approach (and the presentation as a whole) that we've devoted a separate chapter to this subject. See Chapter 6.

Preempt Certain Common Objections. Certain objections (we'll get to the general topic of objections and how to handle them later) arise frequently at various stages of your presentation. Since we know this from experience, we can preempt them *before* they come up. Some objections pertaining to your credibility will be preempted by your handling of the credibility factor (as discussed in Chapter 6). Certain other objections relating to *providing references* are discussed in Chapter 7 (along with techniques for preempting them).

However, there are still other very common objections that will arise with considerable frequency, and you must attempt to preempt them. For example, customers will want to know about things like your company's warranty and service policies. Customers will ask, "How well do you people stand behind your alarm system?"

Questions

The questions segment of your presentation will provide you with a great deal of information on several important subjects—information that will give you the vital insights you need in order to be able to *guide* the prospects to a positive decision. Let's see what we'll learn from our questions and what the knowledge helps us to accomplish.

Establish Fear Index. We'll talk more about the fear index in Chapter 10. We'll also talk about what it *really is* that motivates prospects to make positive decisions about investing in security.

Before we can sell people our solution(s) to the security problem, we have to determine to what extent *they* believe they *have* a problem. If they don't believe they have a problem, they're certainly not going to buy our solutions.

If they have a *casual* concern about their vulnerability, then we know we have to "crank up" their fear index. If they're merely intellectually curious, you will not sell them. In the chapter on buying motives (Chapter 10), we get into this subject in depth, and we provide you with a great many of the actual questions to ask and subjects to discuss.

Establish Alarm Knowledge. It is very important that we know just how much our prospects actually know about alarm systems. Just what are their perceptions of how alarm systems function? Are they totally ignorant, or do they have some frame of reference? Do they know friends, neighbors or relatives who have alarm systems? What kinds are they? Based upon this information, we then know how these perceptions relate to the particular system we are selling.

We know at just what technical level of description and explanation we have to operate. We know the advantages we have to point out (our system's versus those of the system or systems with which they are familiar). We also learn any misconceptions about alarm systems the prospects might have, and we can then set about correcting these.

Establish Price Perceptions. These questions provide us with insights into what the prospects think adequate security/protection would cost, should cost, might cost. If they (as laymen) have an idea that professional quality protection can be bought for their home for $399, we then know that we must effectively convince them that this cannot be done. We have to show them the reasons why acceptable levels of protection can only be obtained within certain price ranges.

Establish Budget Perceptions. We ask the prospects questions about what price level they would be comfortable with. For example, what could their budget accommodate? What price range might be affordable?

By obtaining this information, we can zero in later on proposing a system to them that they can, or are willing to, afford. Obviously, it makes little sense to attempt to sell a $2,200 security system to someone who really might have to stretch to afford a $950 system. Attempting this would be futile and frustrating. By establishing the budget level, we can eliminate the waste of time and the ultimate rejection that will result from attempting to sell a system that is totally out of the prospect's price range.

Establish Buying Motives. It is important for us to know what the specific buying motives of the prospects are. Through this knowledge, we find out which points/subjects to place more emphasis on as we progress through the presentation.

Is the wife nervous because her husband is away periodically? Is he concerned about her being alone during the day? Do they both work, leaving the home unoccupied for long periods? Are they concerned about young children being alone? Have homes in the neighborhood been burglarized?

These are the kinds of questions we should be asking in our attempt to target any *specific* buying motives other than a general, overall concern or fear.

Establishing General and Specific Problems

Obviously, there is no chance of selling people a solution to a problem they don't believe they have. So it is vital that we make certain that they realize they have a problem.

If people don't believe a burglary can—and probably will—

happen to them, then they won't invest in a burglar alarm. In order to make them aware of the dangers to which they are potentially vulnerable, we have to discuss the crime/burglary problem with them and point up the dangers.

General Crime Discussion. Here, we discuss the crime/burglary problem generally and show newspaper crime clippings (see Chapter 10).

Localized Crime Discussion/Examples. A general crime discussion is fine, but it's still general and will not really move prospects. It's only the *beginning* of an awareness of the potential dangers.

What *will* really move them is a discussion with examples of crime in the immediate area of the prospect's home. Here is where we begin to make them perceive the dangers as very real, very possible, and very close to home. Again, we discuss this topic in greater detail in Chapter 10.

Offering a Solution

We've now made the prospects aware that there is a very real crime problem. They have the problem, and we now have an opportunity to offer them a solution to that problem: the security system we sell.

Demonstration. At this point we make a dramatic demonstration of the system. See Chapter 9.

The Security Survey. After the demonstration, we ask the prospects to accompany us as we take a tour of the house and complete the security survey. The importance of and the reasons for the survey are discussed in Chapter 8.

Closing

After completing the security survey, we are in a position to compute the price of the system quickly (the total investment) and proceed to *close*—that is, simply asking for the order. We discuss closing in detail in Chapter 12.

However, just because we've come to the end of our presentation and are asking for the order doesn't mean that the prospects will automatically say "Where do I sign?" Of course not! They'll have questions. They'll have questions that take the form of objections. They'll raise objections in order to keep from making a decision, which everybody hates to make!

Before we can close, we must handle the objection(s). And after handling the objection(s), we'll try to close again—and again, if necessary. We discuss objections and a technique for handling them in Chapter 11.

6

Establishing Credibility

You Never Get a Second Chance to Make a First Impression

THE SPECIAL IMPORTANCE OF CREDIBILITY IN THE ALARM BUSINESS

Webster's Dictionary defines the word *credibility* as "worthy of belief"—"worthy of confidence; reliability." Of course, this is what we must achieve before we can even begin to expect that our prospects will be moved (by our presentation) to make a positive decision to invest in protection for their loved ones and their homes. We must get to the point, and we must get there immediately upon beginning our presentation, at which the prospects perceive us (and the company we represent) as being worthy of confidence and worthy of belief.

Before we can expect prospects to entrust us with the lives of their loved ones, they have to perceive us as people who are capable and worthy of such a responsibility. Establishing your credibility and that of your company is of special importance in the alarm business because, as opposed to most consumer goods businesses, there are *no* established brand names in which there is *built-in* credibility.

Decades of consumer advertising and consumer awareness have already established credibility for General Electric refrigerators, Carrier air conditioning, Sunbeam mixers, and thousands of other similar consumer products *and services also* (i.e., Culligan soft water, H & R Block, etc.). But, in *our* business, we are all, frankly, Brand X. We are, for all intents and purposes, an unknown quantity as we step on stage in front of our prospects.

So, we *must* create and communicate to the prospect *on-the-spot,* sufficient credibility so they will respect and trust us. Everyone knows RCA, Hoover, Pontiac, and so on, but who has ever heard of the ABC Alarm Co.? Don't delude yourself into thinking that because your company is a major commercial alarm company, with some small name recognition in the business community (and even this is doubtful), that consumer prospects will have any familiarity with your name. It's just not true. Thus, you'll have to quickly establish the following facts.

WHO ARE YOU?

What are your credentials for traveling under the title "Security Consultant" and expecting prospects to take your advice regarding the protection of their loved ones and their homes? The more years in the alarm business, obviously, the more stress you can place on your experience.

If you're a relative *newcomer* to the alarm industry or a brand new recruit, then you would *make no mention of your short time in the field* but should concentrate on the strengths of the company you represent. You should indicate to the prospects that you have been *trained* in all aspects of home protection to be a *professional* security consultant.

Your purpose here is to make the prospects as confident in your diagnosis *of,* and prescription *for,* their security ills as they would be if their doctor diagnosed and prescribed for a bodily illness.

WHO IS YOUR COMPANY?

In most cases where consumers make a buying decision, they are dealing with a known quantity such as a major department store or a retailing giant (Sears, JC Penney, K-Mart, or any of hundreds of other equally well-known sellers). In many cases, also, they are buying from a seller from whom they have bought before.

But they don't know you or your company (in the vast majority of cases). And since 98% of the homes in this country have no burglar alarm, they obviously haven't bought from you before. We also know that in almost every case, they know little or nothing about alarm systems.

Obviously then, in light of these circumstances, it is crucial that we establish the credibility of the company we represent. After all, we want the prospects to make a positive decision about a product/service with no brand name, about which they know little or nothing, involving a significant amount of money, from a company they have never heard of and from a salesperson they have never seen before.

From all of this, it should be clear to anyone and everyone that you must quickly communicate to your prospects a number of things concerning your company that will make the prospects *perceive* your company as one worthy of entrusting the lives of their families and the safety of their possessions.

Length of Time in Business

If your company has some length of time in the business, you should make certain the prospect knows this. Obviously, a point is made by this insofar as companies don't stay in business very long if they sell inferior or troublesome products or if they do business illegally or unethically. If your company has been around for some significant period of time, the implication is that "they must be doing something right."

Conversely, if your company is new, you should avoid mentioning this fact. Many prospects resist doing business with new companies because they may perceive them as inexperienced and/or "not here to stay."

Size

If you represent a large alarm company, you should make certain that the prospects are aware of your size; for example, "Mr. and Mrs. Jones, ABC Alarm Corporation is the largest (or "one of the largest" or "among the largest,") alarm company in this area" (or city, county, state, country). If you have this advantage, make sure you use it.

It is, ironically, the human condition to equate bigness with betterness. Even more ironically, we all know that this is *not* the case. Bigger is not always better, but a significant segment of the human race believes this to be the case.

If you're not the largest alarm company, or one of the largest, or among the largest, you may be able to say truthfully that you're "the fastest growing," "one of the fastest growing," or "among the fastest growing." In fact, you should make the point that your company is growing so rapidly because the quality of your work is becoming widely known.

Don't neglect the opportunities to make these points to your prospects. It only takes a second or two, but it's worth its weight in gold in terms of the credibility points you can make.

What if you're not the largest or among the largest? And what if you're not the fastest growing? What if you're a small, one- or two-man operation? Well, now you must take the opposite approach, and tell your prospects that you're "the boss," "the owner," or "one of the owners" and that that's a "plus for you, Mr. and Mrs. Prospect, because that means that I'm *personally involved* in each and every alarm installation." "You have no 'hired hands' involved in protecting your family. You have the owner—the boss—and you can be comfortable that I'm taking the same personal interest in your installation as I would be if I were protecting my own family's safety and security."

"You're not doing business, Mr. and Mrs. Prospect, with

someone who is just a number with some big, impersonal corporation. Nor are *you* just a number to me. You can be assured, Mr. and Mrs. Prospect, that you are important to the future of my business. I need the good word-of-mouth recommendations I'll get from doing a supersatisfactory job for you folks."

The point we're making to you here is that whether you're large or small, fast growing or status quo, there are opportunities for you to score *credibility points*. Whatever "breed of cat" you are, you critically need to have them like you/trust you.

Affiliations with Organizations

Here again (for right or for wrong) prospects equate the fact that your company belongs to various organizations with being reputable and/or ethical. The psychology is that if you were not reputable and/or ethical, these organizations would not allow your company to be a member. This doesn't happen to be the case, but the public perceives it that way.

You should make certain that you communicate to your prospects the fact that you *do* belong to the local, regional, or state Burglar Alarm Association in your area; the National Burglar/Fire Alarm Association; the Better Business Bureau in your town or area; and any consumer organizations that may be active in your area.

If your company does *not* belong to these groups, you should attempt to persuade management to join. If *you* are an owner or a manager, you would be well advised to become a member of some or all of these groups. The tiny amount of money such memberships cost will be returned many times over in terms of the credibility it will help your company obtain.

Special Capabilities

Every company (including yours) has special capabilities, special attributes, special strengths. Even though the particular capabilities you mention may not have any application for the particular prospects you're talking to, there is a prestige factor involved in telling about, for example, the closed-circuit TV security systems your company installs.

Somehow the capability to install complicated, high-security systems reassures the prospects that your company could certainly do one heck of a job in protecting their home—a comparatively simple installation. And of course, that's the implication. In essence: "My goodness, if we are experts in installing security protection for 100,000-square-foot manufacturing facilities, we would certainly not have even a hint of a problem doing a simple ranch home."

Well-Known Commercial Installations

Tell your prospects about any well-known buildings or businesses your company has protected. Tell them about any high-security installations you've done, such as jewelry stores, banks, savings and loan associations, fur salons, and so forth.

Again, not only is there a prestige factor in having done this kind of installation, there is also the implication that if you're capable enough to protect the Rogers & Co. Jewelry Store or the First Savings and Loan Association, you certainly should find a two-story colonial house "duck soup."

CREDIBILITY TOOLS

You should *create* and/or *obtain* an arsenal of credibility tools. Become a collector!

You should have a loose-leaf binder full of copies or photocopies of the membership certificates in the various organizations of which your company (or you) is a member (e.g., Better Business Bureau).

You should collect (and ask for) testimonial letters from satisfied customers. In the case of testimonials from residential customers, you should black out the customer's name. The point here is that it would be a violation of security to show their names and addresses.

You should get letters from customers (and the police, if

possible) in any instance where one of your company's systems prevents a burglary, or is instrumental in apprehending a burglar.

If you have prestige and/or high-security installations, try to get testimonial letters.

If you can't get such letters, you can photocopy the invoices on which these installations were billed. If you do this, make certain to black out the price and any other private information, and showing just the name of the jewelry store, bank, or whatever the case may be.

As your length of time in the field grows longer, your looseleaf binder should grow thicker and thicker. Thus, your credibility arsenal will grow more powerful, as will your confidence.

Ask the police for recommendations. In many cases they will be reluctant to give them, but so what? If you give enough demonstrations of your capabilities and reliability to enough police personnel, you'll find some who will be willing to give you recommendation letters. Here again, it's the numbers game: ask *enough* policemen and you'll achieve your objective. *Firemen* should also be on your target list for recommendation or testimonial letters.

If your company has advertised from time to time on TV, radio, or in the major area newspapers, you should mention this fact: "I'm sure you folks have seen and heard our advertisements on KDKA-TV or WOR radio or in the *Chicago Tribune.*" This is a significant credibility item. Here again, for right or for wrong, a tremendously high percentage of the public equates media advertising with responsible business practice, and you should utilize this tool.

While advertising may not be part of your responsibility, you might discuss the possibility of your company doing a *token* amount of media advertising just so you and your sales associates can benefit from the credibility points that will be derived from such advertising. Even *single* appearances in each advertising medium (TV, radio, newspaper) will be productive in that you can now ask prospects if they've seen your ads.

Don't, however, fool yourself into thinking media advertising will generate leads for you. It won't. We discuss this in a later chapter.

APPEARANCE AS A CREDIBILITY FACTOR

To the prospects you meet every day, you *are* what you *appear* to be. If you *appear* to be a capable professional, they will *perceive* you as such. Certainly, you must actually "know your stuff," but it's equally important to *look* as if you know your stuff.

You must dress and look like a person from whom people would feel comfortable taking advice and guidance. You should look the same way your prospects' lawyer, doctor, or accountant looks. After all, you're just one more of the professionals in their lives from whom they take counsel. Remember, *you never get a second chance to make a first impression.*

Rather than go into a lengthy discussion of the specifics of how you should dress when in the field, we strongly recommend a very highly regarded book, "Dress for Success," by John T. Molloy. The book is available in paperback and is published by Warner Books. This book (over 1 million copies sold) is required reading in hundreds of major U.S. corporations.

If you *feel* that you look successful and professional, you'll soon begin to *act* that way, and this will reflect back to you in the way others *perceive* you and *behave toward you.* It's a self-fulfilling prophecy, one that will prove itself worthwhile with your next W-2 form—the ultimate report card.

POINTS TO REMEMBER

Think about the following points discussed in this chapter. Discuss them with your colleagues; talk about them with your management.

- The special importance of credibility in the alarm business (like me/trust me)
- Who are you (credentials)?
- Who is your company?
- Length of time in business
- Size
- Affiliations with organizations

Establishing Credibility

Special capabilities

Well-known commercial installations

How to capitalize on being large

How to capitalize on being small

Credibility tools

Appearance as a credibility factor

7

Eliminating the Reference "Stall"

The One-Call Close and Other Serious Tactics

Before talking about references, we should comment on the philosophy of the *one-call close*. What do we mean by the term, and what are the implications of this philosophy as it relates to our business, our earnings, in fact, our whole future? The term *one-call close* seems to be simple, self-explanatory, and yet, it's so widely misunderstood and poorly implemented.

Obviously, the term means closing the sale on one sales call—the *first* sales call. This eliminates the need to return to a prospect's home to complete (or attempt to complete) the sale. Sounds simple, doesn't it? And yet, especially in our business, so few salespeople truly realize the critical need to tailor their every action—their every word—toward the ultimate objective: the first-call close.

In our program, we have created a structured presentation (Chapter 5) that is geared to the one-call close. If you use this presentation, *you* will automatically be geared toward the one-call close.

Make no mistake about it, we are in a one-call close business. Over the course of a selling career in the home alarm business, you'll learn that if you don't sell 'em on the first call, in the vast majority of cases you won't sell 'em at all. (This happens to be true not only in the home alarm business but also in practically all in-the-home sales businesses.) In most professionally run sales forces (again, not only in our industry but also in other in-the-home businesses) salespeople are *directed not to go back.*

What makes us believe that, if the prospects weren't moved to make a positive decision on our first presentation, they will change and do it on the second? The only thing that happens in the interim between a first call and the return visit is that the prospects have an opportunity to create all the negative reasons why they *shouldn't* spend the money. All of the emotional ingredients the effective salesperson wove into his initial presentation *are forgotten,* and he or she has to *start all over again* to build the same atmosphere and instill the same fear into the prospects.

One of the primary reasons (among several) that many commercial alarm companies have been so ineffective in the home alarm market is that they don't understand the concept of the one-call close. They are accustomed to giving the business prospect a proposal, giving him the opportunity to evaluate that proposal, and then coming back a second time to get the prospect's answer or decision.

Many alarm companies have such antiquated pricing formulas that they can't even give the prospect a price on the first call. They have to go back to their office and figure out the price and then come back and *propose* this to the prospect. This is an invitation, not to making a *third* call necessary, but really, to disaster.

Actually, the word *propose* should not even be in our vocabulary. *We should discard any printed forms we have called proposal.* We are not in the business of proposing; we are in the business of helping people make positive decisions. We must use *order forms, not* proposal forms!

And of course, one of the surest ways to "kill" the one-call close is to give references. What? Not give references? You're saying, "Why, I have any number of happy customers who would be glad to recommend my alarm system." Of course you do; we all do!

All of us have references we could easily and proudly give to a prospect. And, of course, if we were to be foolish enough to do this, we've succeeded in giving the prospect an easy way to stall us, to avoid the need to make a decision. We've given him the weapon with which to defeat us.

If we give him references, it would take a week or whatever for him to contact the references, and by that time we're back to "square one." We'd all be better off, and ahead of the game, if we took the time it would take to go back and make a second call and invested it in making a presentation to a fresh prospect.

Yes, certainly we've all had any number of instances where we sold people on a return call and we've all had prospects who *did* call us back, but statistically, over the course of a selling career, you'll find that these are a minute fraction of the total.

If your intent is to sell home alarm systems *in quantity* and *not by accident,* you'll acquire the discipline to avoid the temptation to give references. And, of course, it's totally *unprofessional to give references* and violate the confidential nature of what we're doing. Doctors, dentists, lawyers—these professionals don't give references. It would be unthinkable, not to mention *unethical.*

If you do an effective job of establishing your company's credibility, of making the prospect perceive you as a professional, with integrity, and in making a dramatic demonstration, you will, in most cases, *preempt* the request for references. They will respect you for being a professional and not a peddlar. Let them ask a landscaper, a plumber, or an electrician for references—*but not us!*

WHAT TO SAY TO AVOID GIVING REFERENCES

We've talked about the reasons you critically need to avoid giving references, but now we need the actual words for you to say. If the subject of references comes up and if you appear to be the least bit evasive, you will not only damage your credibility in the prospects' eyes, but also you will make them even more intent on getting references from you and, thus, as we said earlier, effectively kill any chance for the first-call close.

What *do* you say? "Mr. and Mrs. Jackson, I can understand your asking for references. However, it's something we would never dream of doing. I'd love to give them to you because, frankly, after you heard some of the things our customers say about us and the security systems we install, my work would be done for me. Their praise and compliments would very quickly help you to make a positive decision to let us get your family and home protected. But to expose the names and addresses of people whom we're protecting would be a terrible breach of ethics and would be totally unprofessional."

"I don't thing you'd want me to be giving out your name and address to the whole world, once we've gotten this home protected. And, after we *do* get you protected, I have no doubt you'd be happy to give a recommendation to some prospective customer of mine."

"It's no different, really, than your doctor or lawyer. You're probably quite happy with the work they do for you, and you'd probably be glad to recommend them. But I don't think you'd like it (and it wouldn't be ethical or proper) for them to give out your name as a reference. It would violate all the rules of their profession(s)."

"Folks, with the hundreds and hundreds of homes our company has protected, I could give you dozens of names. But I wouldn't even think of violating their security—just as I would never think of violating yours."

"I always chuckle to myself when I think about references in our business—the security business. It wouldn't be very secure, *would it,* if we casually gave out the names of the customers whom we're protecting. Why, even Willie Sutton, the notorious bank robber, had three good names he could use as references."

"Frankly, folks, with all the hundreds of homes my company has protected, I wouldn't give you a bad name if I had one, *would I?*"

Obviously, you wouldn't use all of these statements. You'll pick and choose, and use those with which you feel comfortable. You'll modify them to suit your own style, your own particular personality. Or you'll create your own way of saying it, using the preceding statements for direction. But whatever you do, say it. Say it firmly, professionally, with *no* trace or hint of *evasiveness!*

One question you probably have is: What do I do if they absolutely *insist on* getting *references?* In such a case, a judgment call is required on your part. You can't very well now quickly provide them with references, after you've just "been on your high horse" about how giving references in a breach of security; this would make you look unprofessional (doing what you just said you wouldn't dream of doing) and unreliable, and you would lose considerable credibility.

You must now "stand pat" and say, "I can understand your wanting the references, Mr. Jackson, but frankly, I would rather lose the sale than breach the security of our other customers. I would hope that you'd respect us for our ethics and our professionalism in refusing to endanger our other customers."

"And think carefully, Mr. Jackson, about any security company that *would* breach their customers' security, because they're just as likely to breach your security. That's a distinct possibility, *isn't it?*"

In *many* cases, this strategy will be effective, and the prospect will agree, if you employ the tactic with *sincerity* and in a firm but nonargumentative way.

In some cases, however, the prospect will not agree. It will come down to no references, no sale. In such cases, if you feel or sense that the strong request for references is legitimate rather than a "stall," either to keep from making a decision or just a ploy to get rid of you, you must now say, "Mr. Jackson, I can see that references are an important point with you, and I can understand that." While it's absolutely contrary to what we would normally do, let me do this: let me speak to a couple of our very good customers and *first get their permission* for me to give you their names and phone numbers. With their permission, I'll give you a call tomorrow and provide you with their names."

You can now call the prospects the next day and supply

the references. You can afford to do this on the *following day* because your decision to give the references has put the one-call close out of the question anyhow.

You *must* wait until the next day because it is important that the prospect perceive that you actually did get permission. You must do it this way, even though you could probably have given them the references "off the top of your head."

A word of caution: *Don't make the judgment to give references too often.* You'll pay for it—in lost sales.

Obviously, it takes more courage and discipline to "stand pat," but you'll be rewarded for it over the length of your selling career by a consistent number of first-call closes. Don't deceive yourself by believing that because you close a few sales on the second calls that this is the way to go. It's an illusion, a mirage!

ACTION ITEMS

Practice the words we've given you for handling the reference "stall." *Develop* your own variations of these words until you've found the particular combinations with which you're comfortable, those that suit your personality and personal style.

Try these techniques on your next few sales calls. And don't be concerned if you feel uncomfortable with them. Just like a pair of new shoes, techniques have to be broken in until they feel comfortable. Once you're handling the reference stall effectively, you're on your way to becoming a professional "one-call closer"—*one of the world's highest paid skills.*

8

The Security Survey

You Can't Sell Them a Solution to a Problem They Don't Think They Have

The security survey is one of the most effective tools in the arsenal of the professional security systems salesperson. *Don't be in this business without using it!*

Just as your doctor takes your blood pressure, your temperature, a blood sample and asks a multitude of questions before he would consider diagnosing and then prescribing, professional that he is, so we, as equally professional, must perform a security survey. You'd run very quickly from a doctor who would prescribe a cure before he diagnosed, and who diagnosed before he asked questions and gathered some information about what's bothering you. Of course you would. And so, we, as sales professionals, have to do a security survey. We must do this for several important reasons.

First, it is critical that we be perceived as professionals, gathering data in order to make a *diagnosis* and then *prescribing* the proper security system. How could we do otherwise and expect to be perceived as we need to be perceived?

Second, the information we gather in our security survey is needed so that we can quickly price out the job, to make a quote. Obviously, we can't ask for an order without being able to tell the prospect what the total investment will be. It is vital that we have a pricing formula that will enable us to compute the price of the job in *seconds.* If we take too much time computing the price, we run the danger of allowing negatives to come into the prospect's mind and are in danger of losing all the emotional atmosphere we've worked so hard to build.

At the time of computing the price, it is important that we give the prospects something to read in order to divert their thinking while we are figuring. A company brochure, a piece of literature, an article on crime in the area—anything that is related and that will occupy their minds for the short time it takes us to figure the job.

If we work with some kind of antiquated method of having to take the data back to our office and go through long, involved calculations in order to come up with a price, we have thus succeeded in eliminating all possibilities for the first-call close. We have created the weapon for our own defeat.

A third important function the survey performs is to *reinforce the prospects' feelings of vulnerability.* Simple questions like those you'll find in the sample survey in this chapter, and the prospects' answers to those questions, will point up to them how unsecure their home and their lives really are.

Questions concerning locks, doors, shrubbery, front and back yard lighting; about street lighting, proximity of neighbors, fire extinguishers, fire drills; and dozens of others can serve to generate the realization that their home is an easy, vulnerable target for intrusion, fire, robbery, or worse. *And without this realization—that there is a problem—we have no chance of selling them our solution.*

The survey is a basic sales tool especially for those instances when a prospect tells you he doesn't need a security system. The no-need objection is one a prospect often uses to attempt to *avoid* your presentation *altogether.*

A prospect might tell you that he doesn't need a security system; therefore, he is not interested. You then respond with, "Mr. Prospect, I can understand your lack of interest in a security system because obviously you have not been victimized by a burglar or robber. So why incur the expense for a system? However, you are aware of the rapid rise in break-ins, and I am sure that you know of several that have occurred in this immediate area. And the FBI reports that the odds of a burglar entering your home are increasing rapidly every day."

"Mr. Prospect, if a burglar was planning to break into your home, what would be his chances of succeeding? I can make a security survey now that will enable you to determine how susceptible your home is to intrusion. The survey will take only a few minutes, and there's certainly no obligation." Then proceed to ask for the pertinent information requested in the survey report. *Have the prospect accompany you on your tour, by all means.* This is vital! He helps to sell himself as you make the tour.

TIPS ON COMPLETING A RESIDENTIAL SECURITY SURVEY

Certainly, you remember that one of the most vital phases of the organized selling interview is problem-fixing. A sale, after all, is simply the solution to a problem. If the prospect feels that there is no problem, then *no sale!*

The security survey is designed to bring an aura of professionalism to your sales effort. It represents, effectively done, your professional diagnosis of the security ills that beset your prospect. Therefore, do it right! Let's consider this survey together.

Comments about Residence

"Mr. Prospect, you're fortunate in having neighbors close by. However, I noticed the heavy shrubs by the side of the living room window. They could present a real security problem since your closest street light is halfway up the block. You have, as you know, a classic two-story type. However, the windows opening on the side driveway should always be locked."

"How old is your home, Mr. Prospect? The reason I ask is brought to mind by the locks on this front door; are they the same on all outside doors? Also, notice the outside hinges. Along with the ease with which these locks could be compromised, the location of these hinges could constitute a double hazard."

"Do you have any outside lighting, Mr. Prospect? Good lights are a great deterrent to burglars."

"Do you have any guns or stamp collections, Mr. Prospect? What I mean is, things of outstanding value that you would especially like protected."

"Obviously, Mr. Prospect, no survey of this type would be complete without a thorough consideration of fire protection. How many fire extinguishers do you folks have, etc.?" Hit hard on fire drills in homes with senior citizens and young children. You may be responsible for saving somebody's life.

You should be knowledgeable about some basic information about residential security. This is information that you will note in making your security survey and will *suggest* to a prospective customer *whether or not* he buys a system from you.

Results of the Survey

In presenting the results of your security survey, you can and should be specific about your recommendations and suggestions *in areas other than just the security system you propose.* That's what a professional consultant does.

Even with a security system, a home with no outside lighting, weak door locks, and hinges on the outside of doors that can be removed easily is a prime target for a burglar. Suggest to the family better outside lighting, replacing a lock or two, adding a bolt lock, and cutting back outside shrubbery next to vulnerable windows and doors—the basics.

If the prospects follow your suggestions and *become security conscious,* their total protection will only be that much better. But if they are sloppy about security, no security system will really protect them, and chances are good that they'll become a crime victim.

How can people concerned about the runaway crime problem protect themselves and their loved ones? The most important piece of advice you can offer people concerning security is *be prepared.* Most victims of crime in the home invite the criminal through negligence. Preventive measures are the best deterrent to burglars. Master burglar Robert Earl Barnes, who admitted to 300 burglaries involving property worth $2 million, says "40% of all burglaries occur because the victim is careless."

Home burglaries don't just happen at night. In fact, almost half of all burglaries take place in the daytime, so every family must be alert night and day.

How do you deter a criminal from entering a home? Most important: try to understand the criminal. *Professional* burglars will target in on expensive estates, residences, and luxury apartments. But every home and apartment in every block—s2low, middle, and high income—is fair game for the mass of amateur burglars, intruders, teenage thrill seekers, and vandals. These burglaries in middle America are the reason the crime statistics have soared. Unlike the professional, the amateur is the most dangerous intruder. He's often armed, and because he's nervous, he can be lethally dangerous.

Make an intruder's or burglar's chances of being detected or caught greater at a home, and chances are he'll look for easier pickings next door or down the street. Put yourself in the criminal's shoes. Take a walk down your prospective systems owner's block and around his home. The best crime preventive action: *Make it difficult for the criminal.*

Light More Lights, Have Fewer Burglaries. Give the intruder no place for concealment. At night, the home should be protected with outside lights on all sides of the house where a criminal might find access. If bushes, hedges, or trees alongside the home offer concealment, make sure these areas are revealed by outside lighting. Burglars and intruders look for concealment to give them time to break a window or jimmy a door.

Most homes are protected by a night light at the front entrance, but the rear entrance that the criminal usually seeks and finds easy access to is usually totally neglected. Suggest that lights be added in your security survey.

Make Entry Difficult. Access into a home should be made as difficult as possible. The professional or amateur criminal will look for easy, weak points of entry into a home. So, in making a survey of the home, look for the weak points and *then recommend positive action to protect them.*

Secure Basement Windows. Basement windows are weak security points in most homes and one of the easiest access routes for criminals. Make it difficult for the criminal. Suggest that security bars or heavy metal screening be added to vulnerable windows. If these windows are never opened, suggest the frames be screwed shut from the inside. In most cases, basement windows are seldom seen, so whatever is done to make them security proof won't be an eyesore.

Basement security can be improved further by curtaining all windows and preventing a curious prowler from taking an effortless inventory of the contents of a basement.

Install Good Door Locks. Good locks on doors today, even brand new locks, were designed when burglary and intrusion were only minor and occasional problems. Many doors are protected by spring latch locks that rank amateurs can open with ease in seconds. Good door locks are vital.

While it is essential to replace easily picked or defeated locks with the best jimmy-proof locks available, a family's total security shouldn't be placed solely on them. They are an additional deterrent that may make an intruder seek easier pickings at another home. The determined burglar or intruder will get in by breaking a window or prying a door from its hinges with a jimmy (a small crowbar).

FRONT DOOR

The front door is usually a home's securest entrance. The door itself is probably of heavy construction and well lighted by an outside entrance light that in itself tends to discourage would-be intruders. The front door is probably protected by a pin-tumbler lock of a type difficult to force from the outside but easy to pick. For maximum front door protection suggest the following locks.

Maximum Security Pin-Tumbler Lock. Should have a large rectangular tongue snugly fitted into a deep, strong door frame receiver. When properly installed, this lock is difficult and time consuming for a burglar to force or jimmy.

Welded Chain Door Lock. A welded chain lock on the inside will allow delivery of messages or small parcels without the need to unchain. For proper installation, the lock slide should be mounted above at a slant toward the chain mounting.

Inside Deadbolt Lock. This lock will provide important auxiliary protection and can be mounted either above or below a rectangular pin-tumbler lock. Its heavy rectangular latch forms a bolt between the door frame and the door itself that is almost impossible to force from the outside. They're available with either an outside key lock or inside manual locking knob.

OTHER DOORS AND ENTRANCES

The average burglar spurns the front entrance of a home since that is where he stands the best chance of being defeated. As a rule, he'll head for a back door or a basement window. With this in mind, following are some other areas around the house that should be checked for proper security.

Porch and Patio Doors. Most sliding patios or jalousie doors are installed with factory locks on the inside that are easy to force. A small jimmy can usually force the lock right out of its soft aluminum door frame. Suggest that a metal or wood door bar be added to all sliding glass doors.

Screen and Storm Doors. Always suggest that screen and storm doors be locked at all times. This serves as one more barrier to an intruder.

Garage Doors. Even when a family is home, they should keep their garage door closed and locked. The sight of an empty garage is always an invitation for a prowler.

Back Doors. From a security standpoint, the back door should receive as much, if not more, attention than a front entrance because this is a favorite target for breaking and entering.

Make sure that the back door itself is sound. Check all

hinges to be certain screws aren't rusted or too short for the thickness of the door. If this door already has a bolt or a latch lock, suggest addition of a jimmy-resistant deadbolt lock, mounted on the inside near the top or bottom of the door as an extra safeguard. This is especially important if a portion of the door, or its frame, contains glass.

Good Lighting

Good lighting is one of the best burglar deterrents. If the lighting in the rear of the home doesn't match the outdoor lighting for the rest of the property, suggest that something be done about it—even if it means getting an electrician to install several outside lighting fixtures above doors, or flood lights to illuminate back areas. Tell the family always to replace bulbs promptly when they burn out.

Basement Stair Doors

A sturdy bolt-type lock on the main floor side of the basement stairway is vital.

WINDOWS AND GLASS

Glass creates security problems. Wherever a glass area adjoins a door or window lock, extra protective measures should be considered. To gain quick entry, an intruder will break a small area of glass and reach in to unlock a window or a door. A good idea is to suggest that a second inside bolt lock be installed away from the main door or window lock. This means the would-be intruder will have to break additional glass areas to find the second lock, and that increases his chance of detection.

Double-hung windows are usually locked with a conventional half-moon-shaped sash lock. Make sure the top and bottom sash securely fit and permit the lock halves to mate firmly.

Loose-fitting sash locks can be easily sprung from the outside by pressure exerted on either the top or bottom sash.

The addition of key- or bolt-type window locks on ground floor windows can provide extra protection.

LOCKS, BOLTS, AND KEYS

Most security experts agree that a single outer door lock will not provide adequate protection. For example, any tumbler-type lock with an opening for a key also provides an opening large enough for the lock to be picked. Thus, a combination of two different types of locks is recommended. The following sections describe the most common locks.

Spring-Latch Locks. A popular and convenient lock that snaps closed whenever the bevelled latch seats into a striker plate is the spring-latch lock. However, this lock can be easily forced by inserting through the door frame a thin blade or stiff plastic between the latch and the strike plate. Thus, a spring lock should always be augmented by a second lock of a different type.

Dead-Bolt Locks. The dead-bolt lock features a solid, rectangular latch that forms a bolt between the door frame and the door itself. This lock is difficult to force or to jimmy and is a good choice as an auxiliary to any other key lock. It's available for either key operation or manual interior knob.

Barrel-Type Bolt. A barrel-type bolt serves the same function as the dead-bolt lock but is restricted to manual, interior operation only. It may be used with an inexpensive night-latch, and it is difficult to force if it's well made and installed properly.

Chain Locks. Used as a supplementary interior lock, the welded chain offers a valuable security feature. It permits the user to open a door partially to exchange letters, messages, or small parcels without completely unlocking the door. For proper mounting, the slide should be installed just above and slanting toward the chain.

General Lock Security. Whenever a family has recently taken over possession of a house or apartment, they should change the pin-tumbler locks on all exterior doors. A registered locksmith can usually alter the previous combination. *While you are not in the locksmithing business, you should know a good locksmith and recommend him.*

ENGRAVE VALUABLES

One of the best methods of protecting valuables like TV sets, typewriters, stereo equipment, bikes, and cameras is to suggest engraving a social security number or license plate number on the valuables. A small $10 to $15 engraving tool is available in most hardware stores to accomplish this. If lost or stolen, the valuable property will be difficult to "fence," and if recovered by the police, the valuables may be traced back to the owner.

In conclusion, remember that a security system is a family's last line of defense against burglary and intrusion, not it's first. If security is important to them, a family will follow many of your suggestions regarding lights and locks and will make their security much better. If you can make them properly *security conscious,* selling them a security system *becomes very likely.*

Following is a sample of a security survey form. You may want to modify and/or expand it. Have it reproduced, allowing plenty of space for filling in the blanks. *Use* this survey. It will *help* to produce big dollars for you.

Remember to make serious recommendations to your prospects based upon the results of your survey. Over and above these recommendations, the survey has an even more important function: *it is the bridge over which you make the transition to give them the amount of the total investment for the security system you will prescribe as part of your recommendations*—the further transition to a *close* (asking them for the order).

RESIDENTIAL SECURITY SURVEY & REPORT

Consultant: ______________________ Date: __________

Name of Customer: ______________________________

Owns home ______ Apt. ______

Address of locations surveyed: ______________________

______________________ Zip: __________ Telephone: (   )

Number of persons residing here: ______

1. Comments about residence: ______________________

   Location of nearest home: ______________________

   Street lights? __________ Shrubbery, trees? __________

   Stairwells? __________ Shopping areas? __________

   Other unusual aspects? ______________________

2. General observations and recommendations:

   Locks? (doors & windows) _____ Doors & hinges? _____

   Outside lighting? _____ Expensive collections? _____

   Other: ______________________________

   ____

3. Fire Protection:

   How many extinguishers: ______ What type? ________

   Where placed? ______________________________

   Who services extinguishers? ______________________

   Have fire drills been conducted? ______________________

4. Type, if any, of burglar/fire alarm in house: __________

   ____

5. Previous break-in, burglaries, fire, etc.? ______________

____

6. Any physical harm suffered from #5? ______________
7. Amount of dollar loss from #5: ______________
8. Amount of dollar loss covered by insurance: __________
9. Have any neighbors suffered from burglary, hold-ups, fire, etc.? ______ Where? ______________

   When? ______________________________
10. What about relatives and friends? ______________
11. What is history of burglary, attacks, fire in neighborhood?

    ____

12. Type of home: Ranch ______ Two-story ______

    Bilevel ______ Other ______________
13. Type of construction: Wood frame ______ Brick ______

    Other ______________________
14. Types of windows: Sliding ______ Double-hung ______

    Casement ______ Other ______________
15. Window frame materials: Wood ______ Aluminum ____

    Steel ______ Other ______________

SECURITY CONSULTANT'S RECOMMENDATIONS

16. A. Doors to be protected:

    Front _____ Back _____ Side _____ Basement _____

    Garage _____ Other ______________

B. Windows to be protected (add up and insert number):

Living room _____ Dining room _____ Kitchen _____

Family room _______ Den _______ Bedrooms _______

Other ___________________________________

C. Interior (trap zone) protection: ____________________

D. Should protection be expanded beyond main floor?

___________________________ Why? ______________

____

9

The Demonstration

Avoiding the Hardware Peddlar's Syndrome

THE DEMONSTRATION: WHAT WE WANT IT TO ACCOMPLISH AND SOME OF THE DANGERS TO BE AWARE OF:

Yes, a demonstration of your alarm system is vital. It's essential because the prospect, who knows little or nothing about alarm systems, will want to know *how* it functions and *what* it does (for him) and how it actually protects him.

So we want to show the prospect(s) the only thing that matters: when an intruder attempts to enter the home, the alarm is triggered. We then give an orderly, simple explanation and description of the functions of the system and how it is operated. But we do most of this with the *demo case closed* and out of sight. As long as the demo case remains open on the table, the dangers of getting into complicated (and diversionary) hardware discussions are present.

Do the demo—and make it short and dramatic—and then get the demo case closed and *away* from the table. But it's all done in layman's terms, nontechnical language.

We should make the point again, at this juncture, to make certain that we don't allow ourselves to fall into the trap of becoming hardware peddlars instead of what we really are (or must become)—people who skillfully and professionally motivate other people to invest in the concept of protection for their families and homes *now!*

You see, we don't sell control panels, magnetic switches, motion detectors, digital communicators, and so forth. We don't even sell burglar alarms or security systems; these are just *things. And people don't buy things; they buy ideas and concepts.*

So, what we're really selling is the *concept* of "peace of mind." We're selling the idea of protection for the prospects' loved ones. And by selling *conceptually,* you'll find that you can move people *emotionally* to make positive decisions. Few, if any, prospects will be moved by your description of the multitude of features contained in a control panel or of the features possessed by a digital communicator.

It cannot be stressed strongly enough that once you have made a short, dramatic demonstration of your system, you should immediately close your demonstration case and place it *away* from the area where you and the prospects are sitting. *Then* continue your discussion and/or answer any questions about the system they may have. By doing this, you reduce the risk of getting into a discussion of the hardware, and you can continue with the sequence of your organized sales presentation.

In terms of percentages, the make-up of the effective alarm sales presentation is 10% demonstration and 90% concept. Don't become a victim of the hardware syndrome. Don't attempt to impress the prospects with your vast knowledge of the technology of alarm equipment. They don't know anything about hardware, and they care less. Put it all in layman's language. It's more important to impress the prospects with your sincerity in wanting them to be protected.

Remember, they don't care how much you know; they want to know how much you care!

Another penalty you will pay for allowing yourself to become a hardware peddlar is that you will constantly be bedeviled with price and money problems. When you sell the concept, rather than the gadgets, you will encounter few pricing problems. After all, can you put a price tag on peace of mind and security protection for your loved ones? Not very easily!

An excellent example of the point we're trying to make here is the Electrolux Company. This company sells over $500 million worth of vacuum cleaners each year. That's right, over half a billion dollars *a year*—one company alone, doing more dollar volume than the entire residential alarm industry, at a price eight to ten times higher than other vacuum sweepers sold over the counter in retail stores. In addition, most households *already have* a vacuum sweeper, many have *two* vacuum sweepers, and a surprising number have *three.* How do they do it? Why are they effective?

The answer is that they don't sell vacuum sweepers. That's hardware, and they would have a difficult time selling that hardware to someone who already has one or more of what they're selling, at eight to ten times the price of other vacuum cleaners. Even granting the fact that it's an excellent product, which it is, they would not sell $500 million worth of them a year if they were hardware salespeople.

What *do* they sell? They sell a concept: "More leisure for the lady." They demonstrate the benefits—namely, that their vacuum sweeper is so efficient and effective that it allows the homemaker to keep a house immaculate in a fraction of the time it would normally take, thus freeing up the time for more leisure activities and a change in life style.

Can you put a price tag on the concept of "more leisure for the lady?" Can you put a price tag on "a change in life style?" Not only can you *not* put a price tag on these concepts, but also you can sell your product to people who already have one or more! The bottom line here, of course, is that Electrolux sells a great deal of hardware, at a very substantial profit, and they do it not by selling hardware but by selling concepts.

As someone who intends to be a professional salesperson, you must first understand that you do not sell the features of your security system, you sell the benefits of it. You do not sell burglar alarm hardware, you help people make positive decisions about acquiring peace of mind and investing in protection for their loved ones.

Once you grasp the principle of conceptual selling versus

hardware selling, you'll find that you'll end up selling considerably *more* hardware and avoiding price problems to boot. Ironic, isn't it, that you'll sell more hardware by *not* selling hardware?

Again, we keep our demonstration short but dramatic. We keep our conversation in layman's terms—nontechnical—and when we're through we close our demonstration case and put it far away from where the presentation is taking place. But with all of its simplicity, we manage to cover all of the functions of the system, its capabilities and its advantages, effectively. We want to come across as professionals who know our stuff, and yet we don't deliver a lecture because that would accomplish nothing except to confuse the prospects and get us into the hardware peddlar's trap.

Our verbal explanation (without the demo open) is many times lengthier than the actual physical demonstration. By doing it this way, we attempt to avoid the unimportant (but frequent) questions such as Where would this gadget be placed? What colors does this come in? You mean you're going to put those white things at every one of my windows? Do you have this in walnut? What if the burglar does such-and-such?

If you've been out in the field, you've heard them. You can preempt many or most of them by doing it my way! Save the technical discussions for bull sessions with your colleagues. You should say, "We'll want to protect your doors and *certain* key windows. Since we haven't determined what the exact price will be, we then have the leeway to leave certain windows not covered."

We have the leeway to use a variety of area protection devices, as the money and the situation permit. In essence, we then have some flexibility in terms of the specific components of the system.

We make a strong point when we state, "This system works almost identically to those systems you've noticed protecting banks, jewelry stores, and other high-security locations." In fact, we could go further and say, "I figure if it's good enough to protect my bank, it's good enough to protect my home, and it's the kind of protection I feel proud to offer you."

The point is very potent, incidentally, whenever the subject of some over-the-counter or plug-in type of alarm is brought up. In such instances we should say, "I don't think you would find them using such a device to protect themselves down at the First National Bank, would they?"

ACTION ITEMS

Practice, drill, and rehearse a demonstration of the system your company sells. Once you're satisfied that it is *simple* yet *dramatic* and that it illustrates the effectiveness of your particular system, "lock it in"; that is, do it the same way every time.

Certain aspects of security systems are particularly effective in impressing prospects, things like hand-held panic buttons, etc. These are the "sizzle"—the things that help close sales.

Let the prospects participate in the demonstration. This is not only effective but also makes them comfortable that they don't need college degrees in engineering in order to operate the system. This is important because many times prospects will say no not because they don't want the protection but because they are afraid the system will be too complicated for them to operate but are embarrassed to admit this.

Record your presentation on an audio cassette, then listen to it and be hard on yourself. Take notes while you're listening. And then critique yourself on the following points.

- Is there too much technical "lingo" in your presentation? If so, go back through it and eliminate this.
- Is the demo too long? If so, then shorten it.
- Do you cover all the dramatic benefits, the "sizzle"? If not, include them, where they fit.
- Do you sound professional? As if you really "know your stuff?"
- Does the system you are demonstrating sound like it would be effective in preventing burglary, yet easy to live with and operate?

In the final analysis, *honestly*, would you buy the system for your own home? Or for your own mother and father?

Try doing the segment of the demonstration in which the noisemaker is triggered (either by simulated intrusion or panic function) in the dark. Notice how much drama is injected into the demonstration by doing this.

At the same time, in the dark, you set the stage for this drama. "It's 2 A.M. You and Mrs. Prospect are sound asleep. It's dark and quiet and this burglar is going to attempt to enter your house to steal, to rob, to assault, or worse. But our security system detects him, and all hell breaks loose (trigger the alarm).

Let the noisemaker sound for 30 to 45 seconds (this sounds like an eternity in someone's kitchen). Shut it off and then ask the critical question: "Mr. Prospect, Mrs. Prospect, what do you think this creep, this animal, will do when all this happens?" Mr. Prospect replies, "Why, my goodness, he would soon be half a mile down the road, in a flash." Or words to that effect. Of course, that's what the prospect will say! You then proceed to make a transition to your survey and then on to a close.

SUMMARY

Sell the concept (peace of mind), not the hardware.

Sell the benefits (peace of mind), not the features.

People don't buy *things;* they buy ideas/concepts.

They don't care how much you know; they want to know how much you *care.*

Hardware peddlars have price problems, but you can't put a price tag on peace of mind.

You will sell more hardware by *not* selling hardware.

The effective presentation is 10% demonstration, 90% concept.

You are there to communicate the concept and to motivate the positive decision, not to deliver a technical lecture.

10

Buying Motives

The Heart Does Things of Which the Mind Knows Not

We would be hard pressed to think of anything we've ever bought that wasn't bought for one or a combination of the following motives:

Pride,

Profit,

Fear,

Love,

Need.

Well, we know that everyone *needs* an alarm system! But since less than 2% of the homes in this country have one, it follows that 98% of the remaining potential buyers don't really think *they* need one. It further follows, then, that we can't depend on the *need* motive if we are to sell alarm systems in quantity.

The bottom line is that most people really don't think they

are potential intrusion victims. If they did, they'd run down to Sears, Roebuck to buy an alarm system. But, as we said earlier, they don't and they never will! It remains for skilled professionals (like us) to motivate them to do what they *should* do.

To do this, we have to move them emotionally. They will not make a positive decision to invest in an alarm system on an intellectual basis. No, they buy emotionally and then rationalize the purchase intellectually—*after* they own it. William Shakespeare said it all, when he said, "The heart does things that the mind knows not."

If you will refer back to Chapter 3, "The Organized Sales Presentation," you will see that segment C of the presentation is entitled "Establishing General and Specific *Problems.*" Note also that the next segment of the organized sales presentation, segment D, is called "Offering a Solution." Well, we can't go to segment D if we don't properly communicate segment C, can we? Stated more clearly: There's no point in offering solutions to problems people don't think they have! You *can't* sell *your* solution to prospects who don't believe they have a problem.

Thus, the crime discussion is meant to establish in the prospects' minds the fact that not only is there a *general* problem (burglary danger) but also a very specific problem (the likelihood that it will happen to *them*).

Establishing the specific problem (in relation to your prospects) is making them realize that "It *can* happen to me!" If your prospects cannot be made to feel that they are truly in danger, that it *can* happen to them, they will *not* make a positive decision to protect themselves via the purchase of a burglar alarm system.

Establishing a general problem is easy. Our work is done for us in everyday's newpaper, the radio news, TV news, magazines, and every other one of the mass media. No day passes without a multitude of crime stories.

The FBI *Uniform Crime Reports* indicated that there were over 3 million burglaries in 1982. Succeeding years will have that many or more. However, since the U.S. Census Bureau's "Victimization Study" indicates that only about one out of three crimes is *reported,* it would then appear that there were probably closer to 9 million burglaries rather than 3 million burglaries. What a horrendous number! We'd have to dig deeper into the figures to find out how many of those burglaries also involved confrontation with the victims—for example, rape, assault, and in some cases, even murder.

The U.S. Bureau of Justice statistics report that about 25 million U.S. households were touched by crime in 1982. That's about 30% of the total number of U.S. homes. This same government agency reports that U.S. families are *more likely* to have a member attacked in a serious crime than to suffer from a residential fire or an automobile accident.

It's ironic, isn't it, that citizens would not consider being without insurance to protect themselves against fire or auto accidents but neglect to protect themselves against crime, which is even more likely to happen to them. An even more ironic difference is that fire insurance or auto insurance doesn't really protect them. They merely get paid for their loss *after* the catastrophe has happened.

Security systems are the finest kind of insurance because they prevent the catastrophe from happening altogether! Burglar alarms are better than life insurance, where you have to die to obtain the benefits. Burglar alarms are *living insurance;* they save lives while the owner(s) is still here.

We've been talking about statistics from the FBI, the Department of Justice, etc., but, you know, the funny thing about statistics is they don't move people to action. Statistics are cold, unemotional numbers to which people will react intellectually and then proceed to do nothing. The infamous dictator Joseph Stalin said, "One death is a tragedy, but a million deaths is a statistic."

There is a relevance in this statement for us in the security business. If statistics (the general problem) could move citizens to action, as horrible as the statistics are, citizens would have long ago raced down to their nearest retail store, shouting, "Where do you display the burglar alarms?" But, of course, they haven't done this. They aren't doing it, and they never will!

They never will because, like life insurance, burglar alarms are not bought—they're sold! They are sold by professional salespeople who are able to take the general problem and communicate it in terms of its being a specific problem—not a problem that's statistical and relates to the general populace, but a problem that relates to the specific prospect, in front of whom the salesperson is sitting.

Every effective salesperson will scrutinize everyday's newspapers and clip those stories that relate to crime and burglary in all of the various sections of the area(s) in which he or she operates. The effective salesperson will have a briefcase full of clippings, so that no matter what part of town the prospects live in, he will be able to shown them actual examples of crimes and burglaries that have taken place in their immediate proximity, in their neighborhood.

By doing this, the salesperson brings the chances close to the prospect's home where it becomes not a general problem but a specific one. Show the prospect half-a-dozen burglaries in his neighborhood, and he is on the way to believing that "maybe it *can* happen to me." And once he starts believing this, we have an opportunity to sell him a security system. If he begins to believe that he personally has a problem, we have "a shot" at selling him *our* solution.

Remember, the crime discussion comes in that part of your organized sales presentation *after* the questions (section B of the presentation). This means that you are having this crime discussion armed with a great deal of information obtained via your questions. And from the answers to those questions, you'll know just how lengthy you should make the crime discussion. You'll know where the prospects' fear index is.

If they are really nervous, frightened—if they mention break-ins in their neighborhood, if they recount criminal incidents in their area—you may find that a very brief crime discussion is needed. But if they don't tell you of these incidents, if they appear to be simply intellectually curious (you'll know where they stand from your questions), then you should have a lengthier crime discussion. You should show them as many localized examples of criminal incidents in their neighborhood as you have in your arsenal.

Don't be afraid to show them the most horrible incidents such as citizens coming home, surprising burglars in the process of burglarizing the home, and being assaulted or murdered. Don't be reluctant to do this because:

- You're not showing them fiction; these things really happened to real people.
- What you want them to do is to prevent this kind of thing from happening to them and/or their loved ones.
- You must take them from being intellectually curious to being emotionally concerned.
- You must increase their fear index.
- You must motivate them to *want* to do what they *should* do.

Whenever you are reluctant to raise some prospect's fear index, close your eyes for a brief second and visualize the scene at a funeral home where a grieving family is gathered around the coffin of a dead loved one murdered by a surprised burglar. And, then, think further about how that scene could have been avoided if someone (you, perhaps) had motivated that family to get themselves protected.

Just remember that every time you motivate some prospect to stop procrastinating and get themselves and their loved ones protected, you're performing a public service, one that the law enforcement institutions have been utterly incapable of doing. You're a life-saver! And you should never forget it.

ACTION ITEMS

Be sure to read *all* of the newspapers in your town every day. Set up a file, divided into sections for each part of town and each neighborhood, and clip each day's stories about criminal incidents in those areas. In this way, you can quickly and conveniently refer to the appropriate file relating to the part of town where you have appointments to make presentations each day.

Become a voracious reader so that you literally become an expert on crime and burglary. Read the *Wall Street Journal, The New York Times*, etc., both of which frequently carry excellent articles and information on crime topics. Get a subscription to *Time* magazine or *Newsweek* magazine, both of which frequently carry articles on crime. Clip these stories.

Equip yourself mentally to carry on the same kind of discussion on burglary/crime that your doctor might carry on concerning heart disease. Why not, you're both professionals, aren't you?

BUT BUT BUT
BUT BUT BUT
TUBUTUBUTUBU
BUTTA BUTTA BUT
BUT BUT BUT
BUT BUT BUT
BUT BUT

11

Answering Objections

Learning a New Set of 3 Rs

The ability to handle objections is probably the single *most critical segment* of any sales presentation because, obviously, if you can't overcome the objections, you can't get to the close. Instead of dreading objections, we should welcome them because people who don't object aren't seriously involved or interested. They won't buy because they don't care enough to challenge the price, ask for more proof, or question their need to own the security system you're offering.

What is an objection? It's a statement by your prospect(s) that he, she or they want to know more. But, of course, objections never come out sounding like a polite request for more information. Prospects usually are sincerely objecting; they don't realize they're just asking for more information. It is our job, as professional salespeople, to know this and to know what to do about it.

Before we get further into the subject of objections, let's talk about something called a *condition.* A condition (as opposed to an objection) is an actual, real reason for the prospect being unable to go ahead. We can't overcome conditions as we do objections, so it's important that we try to recognize them.

One of the primary reasons we keep qualifying and asking questions is to attempt to determine if there are any conditions that make the sale impossible. If there are—*absolutely* no money for example—we have to be able to withdraw graciously from the situation because it's a waste of time.

We can do great damage to our enthusiasm if we attempt to overcome a *condition* as if it were an *objection*. It can't be done, and it's pointless to continue. Let us quote here two paragraphs (on the subject of *conditions*) from a great book on professional salesmanship by one of the greatest sales trainers in the United States, Tom Hopkins, president of Champions Unlimited:

> A condition is a valid reason for not going ahead. It's not an objection to overcome, it's a total block to the sale that you must accept and walk away from. A Champion is quick to recognize conditions. The major aim of qualifying is to determine whether there are any conditions that make the sale impossible, and further attempts to sell pointless. So the Champion, being an expert qualifier, never destroys his enthusiasm by trying to overcome conditions that can't be overcome.
>
> Even a skilled qualifier may occasionally be well into his selling sequence and then encounter what appears to be a condition. When that happens, treat it like an objection. That is, try to break it down. If it doesn't break down, it's a condition, and you'll need to develop the ability to swallow hard and then quickly and courteously disconnect from that prospect who, you've just discovered, can't buy. Some of us have a severe problem here. If we spend time with people and get into our selling sequence with them, we get emotionally involved to the point where we lose our ability to see the difference between an objection that can be overcome and a condition that can't. In this, sales is like poker. Professional poker players are quick to throw a losing hand, no matter how much they've already bet on it. Average poker players stay in the game and keep on calling bets with hands they know have little chance to win. Instead of accepting a small loss and getting out, they stay in and take a greater loss. As Reinhold Niebuhr said, "Give us serenity to accept what cannot be changed, courage to change what should be changed, and wisdom to distinguish the one from the other."

> Have the wisdom—and sometimes it also takes courage—to pleasantly and quickly withdraw from losing situations.*

Of course, Tom Hopkins couldn't be more correct; conditions are losing situations, and while objections are winning situations, they are the rungs on the ladder of sales success. Let's continue now on the subject of objections.

There are four broad categories of objections, and then there are a whole bunch of objections that are peculiar to our business. Most objections come under one of the following four headings:

1. No need,
2. No interest,
3. No money,
4. No hurry.

For example:

- We don't NEED a burglar alarm.
- We're not INTERESTED in a burglar alarm.
- We can't AFFORD a burglar alarm.
- Call me in a week.
- See me after the first of the year.
- Check back with me after Labor Day.
- We want to think about it.
- My wife and I want to discuss it.
- We've got a watchdog; we don't NEED a burglar alarm.
- I've got a gun; we don't NEED a burglar alarm.
- We have plenty of homeowners insurance.
- We had one alarm company here last Wednesday, and another one coming tomorrow, and we're going to get three quotes.

---

*Reprinted with the permission of Tom Hopkins from *How to Master the Art of Selling*, Champions Unlimited, Inc., 1980.

- Why is your system $1,800 when the system the other guy showed us was only $1,100, and it sure looked a lot like yours?
- I saw a burglar alarm advertised at one of the local electronic stores, and it only cost $89, and all you have to do is plug it into an electrical outlet.

If you've been out in the field selling for a short time, you've already heard many or even all of the preceding objections. Once you've been in the field for a while, you will have encountered all of them.

You must *learn, practice, drill,* and rehearse *your* answers to each and every one of these objections. Many times, three, four, and five of these objections will arise in a specific presentation. If the prospect has four objections and you can only handle three of them effectively, you'll never get any further toward your primary objective—the close.

Before we get into the technique for answering objections, one word of caution: Even though some of the objections you encounter may seem (or even really be) foolish or irrational, don't reject these objections as being anything less than reasonable and intelligent. Remember, your object is not to prove them wrong but to give them the information they *think* they need or *want.* You aren't there to argue. If you do, you'll likely win the argument and lose the sale. Don't risk winning the *logical* battle and losing the *emotional* battle.

Now, for a technique for handling objections. We call it the *3 Rs,* which stands for

- Repeat,
- Reassure,
- Resume.

Once you internalize this technique—that is, make it a part of you, make the use of it just like a reflex action—you've made a major step toward becoming a true professional salesperson.

When the prospect says he can't afford it, you say, "I can understand your saying you can't afford it, Mr. Jackson. With inflation and the cost of living being what it is today, sometimes it's hard to see how a person could afford anything. I'm a family man myself, and I know just what you mean." Note that we *repeated* his objection. It's only courteous to acknowledge the objection; it's rude and insensitive not to.

"But, you know, Mr. Jackson, if I had the gift of prophesy and could tell you that beyond a shadow of a doubt that next Tuesday at 12:07 A.M. some animal, some junkie, would enter your home to steal, to assault, or worse, you'd find the money, wouldn't you? Of course, you would."

"Well, Mr. Jackson, Mrs. Jackson, unfortunately, I don't have the gift of prophesy and I can't tell you, beyond a shadow of a doubt, that this will happen *next* Tuesday at 12:07 A.M. But statistics tell us that there's a very good chance it will happen *some time*. You just can't afford to take the risk of leaving yourself vulnerable. Don't make yourself and this wonderful family of yours a candidate for becoming a crime statistic, maybe a personal tragedy." Note that this is the *second* R, the *reassure* part of the technique. And now we *resume* (the *third* R).

"Let me show you here what this will look like on paper. My company has a deferred investment plan in which, with a modest initial investment and an equally modest monthly investment, you can own this family protection system and have the peace of mind you need and deserve."

We proceed now with the remainder of our organized sales presentation. REMEMBER, *repeat, reassure,* and *resume.* You'll want to begin collecting dramatic, relevant answers to the various objections you'll encounter, answers that fit your personal style and personality and that you'll mold to be used in the 3 Rs technique. You'll also want to keep adding to your repertoire of facts, stories, anecdotes, etc. that will be your arsenal of weapons with which to deal with objections.

Once you are comfortable that you have the ammunition to deal with any conceivable objection that may arise, you'll be amazed at the feelings of confidence you'll have and the smoothness, polish, and effectiveness that will come to your presentations. You'll begin to feel like a pro who methodically, scientifically travels toward a close, rather than an amateur who gets his sales by accident. In the following, we present some examples for using the 3 Rs.

REPEAT Mr. Prospect, I can understand your saying you can't afford it. With the cost of living being what it is today, sometimes it's hard to see how you can afford anything.

REASSURE But isn't it true that the phrase "I can't afford it" is one that we apply to luxuries, not necessities? Somehow, we always seem to be able to afford the things we critically need. This security system may save your life or the life of a loved one. And it's not a question of *if* an intruder will attempt to enter this house; it's a question of *when*. It *will* happen eventually, and the time to defend yourself against that intrusion is *now*. Better a year too early than a day too late!

RESUME Frankly, the only honest way to determine just how vulnerable this home is to intrusion, or worse, is by making a professional security survey of the premises. Certainly, there is no obligation on your part, and I think we'll be able to analyze your potential problems from that survey.

* * * *

REPEAT Mr. Prospect, I can understand your saying that you don't need a burglar alarm. You've lived here for twelve years and haven't had a burglary.

REASSURE Mr. Prospect, do you carry fire insurance on this home? Have you ever had a fire (no)? Then why don't you drop it? You've been paying all these years and never needed it (peace of mind, can't take the chance, etc.). Well, Mr. Prospect, if you rightly consider it wiser and prudent to protect yourself against something that *might* happen, isn't it considerably wiser to protect yourself against something that crime statistics say *will* happen?

RESUME Let me show you how our system can give you the kind of protection you'll someday thank your lucky stars for having. Etc.

* * * *

REPEAT — I can understand your wanting to think it over, Mr. Prospect. That's just being a prudent consumer.

REASSURE — But is there something I haven't fully explained to you? The way the system protects you? The way it operates? Is there a budget problem? Then, really, what is there to think about?

RESUME — Let me show you a deferred payment plan that my company has created. Etc.

* * * *

REPEAT — I can understand you folks wanting to think it over. That's being a good consumer. I respect you for that.

REASSURE — But, folks, the horrendous rise in crime has been frightening, *isn't that true?* And it's obviously going to go on for years, *isn't that true?* The point I'm trying to make, Mr. Prospect, Mrs. Prospect, is that you can think about this for a week, a month, or a year, and you still won't be any closer to making a decision than you are right now. The only difference is that, in the meantime, while you're thinking about it, this home is a candidate for becoming a crime statistic, a personal tragedy.

* * * *

These several examples should give you a good idea of how the repeat-reassure-resume technique for handling objections works. And it does work. It's basic, fundamental.

Do you see, also, how it enables you to make a transition into an attempted close? It's the orderly, natural next place to go to. Answer the objection, then try to close (ask for an order). This may not work. There may be another objection. Answer it, and attempt to close *again.*

Statistics say that the average sale (close) is accomplished on the *fourth or fifth attempt.* So you'd better have plenty of answers in your arsenal. And that brings us to what may be the most important set of exercises in this entire book.

ACTION ITEMS

Take *each* of the objections listed earlier in this chapter, and develop one *or more* logical answers to each objection. Then put *your* answer into the repeat-reassure-resume format, and rehearse it in a *natural, conversational* manner.

Practice, drill, rehearse these answers. Internalize them; make them a part of your reflexes.

Begin to build an arsenal of stories, anecdotes, and answers to objections so that you travel the road "armed to the teeth." You'll be amazed at the confidence this will give you. You'll feel like Superman because you know you can effectively handle *any* and *every* objection they throw at you.

Develop *second* and *third* answers to the same objections so that if they repeat it, you have another rebuttal. Trade answers to objections with other alarm salespeople in your city, at trade shows, conventions, seminars, etc. You'll be amazed at how many new and effective answers you will accumulate.

This is a crucial part of your continuing education as a professional. Remember the 3 Rs:

- Repeat,
- Reassure,
- Resume.

FINISH

12

Closing

The Process of Helping People Make Decisions that Are Good for Them

Before discussing the specifics of closing, we should comment briefly on just what closing is. We have to realize, for openers, that people don't like to *buy*, they like to *own*. Further, people hate to make decisions—even positive decisions they've agreed (throughout the presentation) will provide them with important benefits (peace of mind, security for their loved one, etc.).

So closing (in our business) is the process of helping people make positive decisions that are good for them. The average person can't make decisions about investing in *anything* without help. And the reason so many people need help in making decisions is that they're afraid of making *bad* ones. The famous Tom Hopkins, to whom we referred earlier, said, profoundly, "People love to buy *after* they own." They only have problems with indecision, insecurity, and procrastination *before* they own, and that's why we must become skilled at making people feel as if they already own it and are already enjoying the benefits. That's why we are continually asking questions throughout our organized sales presentation, *getting agreement on minor points throughout.*

For instance, after we've made a dramatic demonstration, we ask, "Well, if he tried to get into this house, and all these bells sounded and lights flashed on, he wouldn't stick around this house very long, *would he*?" You're asking them a question with which they can't help but agree and that enables them to visualize the benefits of the system in *their* specific *house.* "I think you'd all feel a great deal more secure with this kind of system protecting you, *wouldn't you*?"

WHEN DO YOU CLOSE?

By this time, you know that you normally (in the sequence of your organized sales presentation) make your *demonstration,* then tour the house with the prospects to make your *security survey,* and then *quickly* compute a price for the system (the *investment*). This point in the sequence is *critical.*

First of all, while you are doing your calculations to determine the price of the system, you must hand the prospects something to read in order to divert their attention from the threatening nature of what you are doing (calculating the cost). If their attention is not diverted and there is silence, they will have buyers' anxiety. While you're concentrating on your scribbling, the buyers don't have anything to occupy their minds except fear. You should hand them *both* a company brochure, a piece of literature, a crime story—anything to occupy their minds and divert their attention away from what you're doing.

Make sure that you've had your order form *under* your security survey forms *from the beginning of the presentation.* By doing it this way, when you're ready to go into your closing sequence, you don't have to reach into your briefcase to pull out the paperwork and forms. If you reach into a briefcase or fumble with paperwork, the prospect tightens up and is thinking of ways to escape.

At this point—the point at which you tell the prospect(s) the amount of the *total investment* (not the *cost*)—the prospect(s) will react in one *or more* of the following ways:

- They will object—that is, "I can't afford it," "Let us think about it," etc.
- They will give you an opening/opportunity to try a "test close"—that is, "Does your system come with just a plain old off/on key instead of that key-pad?"
- They will transmit buying signals, usually in the form of questions and/or requests, such as "Would you demonstrate that for me again?" "Will you go over that one more time?" They develop a desire for more "technical" information. They talk about things that would happen if they owned your security system.

If they object, you reflexively field the objection (repeat-reassure-resume), and make the transition from there to an *attempted* close such as, possibly, an alternative of choice question: "We're really busy in our installation department, but I think I could arrange to get your installation started on Thursday afternoon, or would Friday morning be more convenient for you, Mr. and Mrs. Prospect?"

If the prospect(s) selects/answers with one or the other of those dates and times, you then simply ask them if they would just *"OK the paperwork, please."*

We should comment here about what you should *always* do *any time* you ask a closing question. What you do here will determine, to a significant degree, your success in closing. *Whenever* you ask a closing question, you must then *be silent. The first person who speaks loses.* This is critical, crucial! Why is it so important to be quiet and not say a word? If you speak first (after asking any kind of closing question), you relieve the pressure on the prospect to speak first, to answer the close and commit him- or herself to buy.

It is very difficult to be quiet after asking a closing question; the silence is thunderous; the twenty or thirty seconds of silence seems like a year. But the skill, courage, and concentration to be still and silent for half a minute is probably the single most vital skill there is in selling. It takes discipline, but the rewards are immense. If you cannot learn this skill, you will destroy close after close after close.

If the prospect does *not* select one or the other of your

alternatives (and you have remained silent), in most cases they will then either ask a question or raise an objection. What has happened, then, is that you tried to close, and it didn't work the first time. You then either handle the new objection, or answer the question, and then proceed to attempt to close *again.*

For instance, if the prospects say (ignoring your question about whether they would like it installed Thursday afternoon or Friday morning), "Does your system come with just a plain off/on key, instead of that key-pad?", you then say, "Do you *want* your system with just a plain off/on key?" When they say "Yes," *they've bought it.* And you simply then make a note of it and ask them to OK the paperwork.

If they *don't* ask you a question but, instead, object—that is, "It makes sense, but we can't afford it"—you then must handle this with your particular rebuttal to the "I can't afford it" objection, and make a transition to *another* attempted close. When you attempt *this* close, you must again ask the closing question, and then be quiet!

How many times must you do this? The answer is as many times as it takes. Many sales will require four and five and more attempts. But, remember, if the prospects have more reasons for not going ahead than you have techniques for helping them to go ahead, they aren't going ahead. It's that simple!

You must learn multiple closes! Here's the sequence:

- Make demonstration,
- Perform survey,
- Compute investment,
- Answer question(s) and/or answer objection(s),
- Attempt to close,
- Answer objection(s),
- Attempt to close,
- Answer objection(s),
- Attempt to close,
- Answer objection(s),
- Attempt to close, etc.

Look for buying signals throughout this process.

Have the courage, knowledge, and discipline

- To ask closing questions (otherwise, you're an order taker with an order taker's income).
- To continue attempting to close, even after multiple objections.
- To understand how to make the transition from repeat-reassure-resume to asking a closing question.
- To be silent after asking this question.
- To start writing quickly after getting what you interpret as a positive decision to go ahead.

Don't hesitate to start writing. The quicker you get the order form written, the quicker you can ask them to "OK the paperwork" (never "sign this contract"). Even if they stop you with "Wait a second, we're not ready to give you an order" or "Aren't you getting ahead of yourself? We haven't decided to buy yet," and so on, your answer is "Oh, I understand that. I'm simply making notes on my paperwork so that my thoughts will be organized and I won't run the risk of forgetting anything." Then you continue answering questions and/or handling objections and going for another closing atttempt.

ACTION ITEMS

Set up a mock presentation and have your hypothetical prospects (maybe fellow alarm salespeople) throw objections (and questions) at you. See if you can not only handle multiple objections but also make smooth transitions to the next closing attempt(s).

Develop your own arsenal of closing questions; for example, "We could probably start installing this system on Tuesdsay afternoon, or would Wednesday morning be more convenient for you?"

Study (like all professionals) closing as a specific technique. We can recommend one of the best books on professional selling ever written, *How to Master the Art of Selling*, by Tom Hopkins, which has two brilliant chapters on closing.

Practice putting the following words on to your statements and questions:

- Doesn't it?
- Couldn't you?
- Can't you?
- Wouldn't you?
- Won't they?
- Isn't it?
- Aren't we?
- Didn't they?
- Don't you agree?
- Haven't they?
- Didn't we?
- Weren't they?
- Wasn't it?

What we are doing here, by using these phrases, is getting agreement on minor points (sometimes major) throughout our presentation.

Never use rejection words (negative) in your presentation. Use go-ahead words (positive).

Never use negative/ rejection words:	*Instead use positive/go-ahead words:*
Cost (or price)	Total investment
Down payment	Initial investment
Monthly payment	Monthly investment
Contract	Paperwork
Buy	Own
Sell	Involve
Sign	OK (or approve)
Pitch	Presentation
Deal	Transaction

13

Creative Prospecting

You've Got to Get 'Em into the Tent before You Can Preach to 'Em

No matter how good your organized sales presentation is, no matter how good you become at handling objections, and no matter how strong you are as a closer, it all becomes academic unless you are consistently and constantly in front of prospects. Since selling is simply a numbers game, you must make your presentation to numbers of prospects each day and each week to make a year! If you are a one-out-of-three closer, you need to make fifteen presentations per week to make five sales. It's as simple as that!

Your company will, no doubt, supply you with a certain number of leads, generated from company-wide lead-generating activities, whatever they may be; and that's great. However, to be utterly realistic, there will *never* be enough company-supplied leads, enough, that is, for you to make the kind of income you want/need. You can depend *only* on yourself to generate enough leads. You must become a dynamic "prospector."

Ask any high-income salesperson in any one-on-one sales business the reason for his success. In almost every case, the answer will be "Prospecting: daily, effective prospecting." The ability and motivation to prospect for leads, it has been estimated, will account for 80% of your selling success.

The endless turnover of protection prospects and suspects demands that you build a good list of qualified prospects on a daily, continual basis. The fact that you had a good sales day or sales week cannot affect your daily prospecting.

Studies concerning problems with prospecting highlight the following shortcomings that keep less successful salespeople from finding and qualifying enough good prospects:

- They don't know prospects exist.
- They don't know where to find prospects.
- They're too lazy to look for prospects.

Every salesperson, even the best, passes up good prospects every day without recognizing them. New neighbors move into homes on the salesperson's own block. Old customers may now be ready for fire-smoke protection. Visitors in a home where a system is being installed aren't asked if a demo can be arranged at their home. Homes next door to installations aren't "clover-leafed" by the salesperson personally. People with whom the salesperson does business—the doctor, the dentist, the gas station owner, the banker, the cleaner, the store owner, etc.—aren't even told the salesperson is in the alarm business. A businessman who decides not to install a system in his office isn't talked to about protection for his home. Every salesperson will miss these prospect opportunities to some degree, but the good salesperson takes nothing for granted; he does more checking, more prodding, more asking, more questioning.

SOURCES FOR SECURITY SYSTEMS PROSPECTS

Present Customers

Good salespeople use present satisfied customers to forge an endless chain of prospects. Customers are flattered when you ask them for the names of friends who should probably also have protection. This request for prospects is particularly timely just after installation has been completed and the customer is in an agreeable mood.

The endless chain method of finding prospects works like the familiar chain letter. Suppose you start with one of your present customers and get leads from them ending up with four more customers. Repeating the process with equal success just fifteen times would add 32,000 customers from prospects started with just one customer! Of course, like the usual chain letter, the system doesn't work perfectly. But would you settle for just 3,200 new customers? Or 320? *How about 32?* However it comes out for *you*, the endless chain technique is basic. Ask your customers for prospect leads and referrals.

Former Business Customers

Whatever your former occupation, if you were in sales, your former business customers are excellent prospects for protection. If you weren't in sales formerly, what about people you used to work with? *Put them on your prospect list.*

Similar Businesses

If you've sold a couple of doctors or lawyers or accountants, use the sale to one type of customer as a wedge to sell a system to other lawyers or doctors. Or, if you've sold a certain type of business protection, use it as a wedge to sell other allied businesses.

Centers of Influence

There are people whose particular position gives them influence over other people, as well as information that can enable you to identify good prospects. Business executives, bankers, police chiefs, fire captains, building managers or real estate management executives, presidents of clubs of all kinds, and civic leaders are examples.

The center of influence, when he is helpful to the salesperson, goes beyond the simple referral. He may make appointments or recommend the salesperson or his position. The center may offer his influence out of pure friendship, or he may expect favors in return. The least he expects is to be kept informed and to be appreciated. *The smart salesperson does these things.*

"Bird Dogs"

Sometimes these are called "junior salesmen" or "spotters." They keep a lookout for prospects for the salesperson, and if a sale is made as a result of the tip, the "bird dog" is compensated.

Route men, service and repair men or women, and other people whose jobs involve meeting the public make good spotters; so do policemen, insurance salespeople, and firemen.

Newspapers

News stories give many leads on burglaries, robberies, and fires and should be followed up quickly.

Directories

The classified telephone directory is a prime source of leads by categories: doctors, lawyers, churches, retailers, etc. The tax assessor's list shows the address and assessed valuation of every property owner in your area.

Tax lists, building permits, voting registration lists, and license lists are just some of the sources that show names, addresses, and other data on general or specific types of prospects.

Personal Contacts

Make up lists of everyone you know who might be a prospect. Make certain they know you're selling home protection. Add to this list as you remember additional friends and acquaintances. Make appointments and presentations to the people on this list. When they're not interested themselves, get referrals from them. In most cases, many of these personal contacts will radiate business for you and will be excellent "bird dogs".

Listening

Be a good listener! Conversation with friends or acquaintances, or even plain eavesdropping, can give you prospect leads. You go to clubs and social gatherings; the people you meet there are potential prospects or represent prospect firms. You must use good taste and avoid open solicitation, of course, and you must be careful not to betray a confidence. But with tact and diplomacy, you can make good use of your listening habits.

Trading Prospects

Look for a noncompetitive salesperson selling home remodelling services, etc., and trade prospects or people's names to whom you've sold a system. Trade lists with him.

Canvasing

There is one other way of prospecting that will never lose its value. Some people call it "cold canvasing"; we call it *"direct approach prospecting."*

As a security salesperson, you are an independent businessperson. As such, it behooves you to introduce yourself to as many people and homeowners as possible. Business people's doors are always open and they are invariably fine people. Most important, they will want to help you.

You owe it to the people in your community to perform as many security surveys as possible. This professional approach to the security business will let the people with whom you desire to do business know that you are not a fly-by-night, that you are concerned about tailoring a security system to their individual needs. Direct approach prospecting keeps you constantly in front of people. It sharpens your skills in handling people and in thinking while on your feet. *You will also find it to be very profitable!*

Remember, the greatest salesperson in the world is unemployed unless he or she is face to face with a prospect!

Cloverleafing

You should *personally* visit homes within a several-block radius of every installation. Tell these people that your company is installing a security system in the home of one of their neighbors and that they should not be concerned when they hear the noisemaker being tested. Present these people with your card and brochure and attempt to set up an appointment that night or in the next day or two to demonstrate your system.

Remember, when cloverleafing (or canvasing), you are *not* selling a burglar alarm, you are selling an appointment to make a presentation/demonstration.

Telephone Prospecting

One of the most valuable tools for you is the telephone. Almost without exception, successful salespeople of all types of goods and services attribute a great deal of their success to the development of good telephone techniques and habits.

Listed are some hints for developing habits that will make telephone time profitable to you:

- Designate a regular time for telephoning each day. Let nothing interfere with it.

- Have the names of people to be called in front of you.
- Use an organized telephone talk. Forget so-called "native" ability.
- Group the calls by area to save time later between appointments.
- *Sell only the appointment.*
- Be *ready* for objections. Don't be thrown by them.
- Pause between dials only to make brief notes on your cards. *Go to the next number fast.*
- *Speak with a smile.*
- Don't argue. *Remember that the objective is only an appointment,* not the sale of a security system.
- Use conviction and enthusiasm.

Never make an appointment more than two days in advance. Schedule times to make three presentations each evening. Be sure to have at least three appointments so that if any don't keep, there will be back-up appointments.

Get a *reverse* street directory. Arrangements for those can be made through your local telephone business office. They are usually made available at a modest monthly rental. If your phone company doesn't publish this directory, go to the public library and look for *Cole's Directory, Hane's Directory,* or *R.L. Polk Directory,* or some other similar directory. These directories list telephone subscribers by address.

Read newspapers and clip out reports on burglaries, assaults, etc. in residential areas. Mail out 20 reply mailers or letters or more every day to the neighbors of the crime victim. Follow up by telephone in three days with the technique suggested. *Do this daily.*

TYPES OF TELEPHONE APPROACHES

Following are approaches to use with a "suspect" to whom you have sent a pre-approach letter.

"Mr. Suspect, my name is Simpson, Bob Simpson, of __ Security Systems. I'm calling about a letter my company sent you a few days ago (pause). I assume you received it (pause). Could I see you this evening at 6:30, or would 7:15 be better (quickly verify address)?"

Your prospect will object in most cases.

For instance:

Objection: I'm not interested.
Answer: I can appreciate that, Mr. Suspect, and while you may not be at all interested in talking about a security system at this time, I would like to meet you. May I see you this evening at 6:15, or would tomorrow at 8:00 be better?

Objection: I'm not in the market.
Answer: Same.

Objection: What is it all about? Tell me over the phone.
Answer: Well, I assume you are not in the market for a security system at this time, are you, Mr. Suspect? Well, I can appreciate that. However, I have access to information that may be valuable to you even if you never buy such a system. I would like to stop by and meet you. Would tonight at 6:00 be convenient, or would 7:30 be better?

Objection: If it's about burglar alarms, it won't do you any good.
Answer: Mr. Suspect, do you say that because you're not in the market for a security system at this time? Is that what you mean? (Yes.) Well, I can appreciate that. However, I have access to information that. . . .

Objection: Mail me the information.
Answer: I will be glad to do that, Mr. Suspect, but its value to you will result from several questions that I know you will want to ask me. Would tomorrow at 6:00 be convenient, or would you prefer 7:30?

Remember, you have only one goal: sell the *appointment!* Do not discuss product, cost, or application. Properly used, this technique will produce consistent leads.

TIPS ON SETTING UP APPOINTMENTS

Radio, TV, or Other Media Inquiries

"Mr. Jones, this is Bill Clark with XYZ Protective Services. I received your card today requesting information on our protective service for your home. May I come right over?"

If prospect can't see you now, ask him, "What time can I see you at home, this afternoon?" If an afternoon appointment can't be set, tell him, "I'll meet you tonight." Then, "I don't want to disturb your supper plans, so I'll be there as soon as you get home. What time will that be?"

Referrals

"Mr. Jones, Mike Ray asked that I call you. I'm Bill Clark with XYZ Protective Services. We recently installed our security system to protect Mr. Ray's home and family. He requested that I show it to you. May I come right over?"

See the preceding example for answers to objections to your coming now.

Follow Up

Personally follow up prior sales. Check your sold cards.

Make it a point personally to see him today. Is he happy with our services? What can you do to offer him superior services? Give him a point of reference to draw referrals from. "I'm sure you have talked to some people at *your office or the club* who need protective service. Give me their names and I will send them a brochure."

Thank You

Personally stop by today's installation and give the customer your "thank you." Get referrals from this customer; this is by far the best lead. "I'm sure you have talked to some people who need our protective services."

Get the names of their neighbors and cloverleaf. Call on these people in person. Explain that we are installing a security system for the Jones family. "When you hear their alarm bell, we are only testing, so don't be alarmed!" Give the prospect a brochure and get the phone number so you can call them to answer any questions they may have. You now have a prospect.

Direct Mail

Following is a series of direct mail letters that you can use in your prospecting activities. You can *revise* them to fit *your* specific situation. Always accompany these with a professional quality piece of literature. Type on plain white paper. An instant printing shop will reproduce these letters on your firm's letterhead.

You should mail a small quantity of small direct mail every day and *systematically follow up by phone* two or at most three days later, using the telephone method described earlier.

You should sign each letter by hand, using a felt-tipped pen. You may even want to use the same felt-tipped pen in doing the salutation: "Dear Mr. Johnson" or "Dear Dave."

SAMPLE FORM LETTERS

For Friends, Neighbors, Acquaintances, Etc.

Dear Mr. Johnson (or Dear Dave),

I'm pleased to announce that I've become associated with XYZ Systems. I am selling professional quality burglar and fire alarm systems to families here.

I'm excited about my new association because we're providing families with needed and vital protection at a *reasonable and affordable cost!*

I realize that most of us say, "It won't happen to me," but unfortunately, statistics prove that it will! With XYZ, at last you can protect *the most important thing in your life—your own family.*

Even if you're not in the market for protection right now, I believe you'll be pleasantly surprised at the reasonable cost of our protection system, and perhaps you'll be able to plan for this vital family protection for the future.

One thing I want to emphasize: There's no obligation on your part whatsoever. Just give me a call to set up a mutually convenient time.

Sincerely,

(Your name)
XYZ Systems

For Burglary or Fire Victims

Dear Mr. and Mrs. ______,

I do not mean to take advantage of your recent misfortune, for I feel certain that you have already been besieged by calls from other security systems sales offices.

I have purposely delayed writing to you in the hope that by now you have had some discussions relating to the installation of a security system in your home. My reason for writing this letter is that I would now like the opportunity to discuss with you the XYZ Security System, the burglar and fire alarm system which has fast become this area's leading name in residential protection.

We are professionals, and we install a professional-quality system.

I would appreciate the opportunity to give you a free home demonstration and estimate. I will take the liberty of calling you within the next few days.

Sincerely,

(Your Name)
(Your Title)

Encl.

For Families in Immediate Area of an Installation

Dear Friend:

One of your neighbors has purchased an XYZ Security System.

Thousands of our security systems have been installed nationally, and we are now in the process of installing one of our alarm systems in your immediate neighborhood.

You may see our installers or hear the electronic alarm bells being checked out. Don't be alarmed, but consider this an invitation to examine an XYZ Security System for protecting your own home and family.

The XYZ alarm system is a professional-grade residential security system. *It can give you quality burglar and fire protection at a price you can afford.*

Please read over the attached literature and accept our invitation to give you a free security survey and a no-obligation estimate.

Just return the reply card, or call me, and I'd welcome the opportunity to show you how XYZ can give you complete protection and peace of mind.

Sincerely,

(Your Name)
(Your Title)

Encl.

For Apartment Dwellers

Dear Mr. and Mrs. ______,

Last week, we installed an XYZ Alarm System in another apartment in your building.

Next ______(Tuesday evening)______ I'm going to be demonstrating this burglar and fire alarm system to other people in your building and would appreciate it if you'd give me the opportunity to show it to you also.

No one wants to be a crime statistic, but unfortunately, none of us knows when a burglar, rapist, or worse might choose us next. With an XYZ System, you at last have a way to protect yourself and your loved ones against intruders and fire—at a price that makes good sense.

The demonstration takes only a few minutes, and I assure you there is no obligation on your part. I'd like to take the liberty of calling you in the next day or two to see if I might show you the system while I'm in your building.

Sincerely,

(Your Name)
(Your Title)

Encl.

For Doctors, Dentists, Etc. at Their Professional Offices

AN IMPORTANT ANNOUNCEMENT TO DOCTORS!

THIS LETTER MAY BE WORTH A LOT TO YOU
IN TERMS OF PEACE OF MIND!

It's about XYZ Security Systems—the revolutionary, patented breakthrough in security systems that may be of particular importance to you.

Our message is short and simple:

1. *If you don't now have an alarm system protecting your professional offices, you should!* Just stop for a moment to consider how a burglary or vandalism could play havoc with your offices and your practice. XYZ can give you *quality protection at a reasonable cost*—an unbeatable combination. We'd be pleased to make a no-obligation security survey of your offices and provide you with a written estimate of what XYZ protection would cost.
2. *If you don't have a home alarm system (guarding against burglary, fire, smoke, and gas), you should!* We'd be happy to provide you with a comprehensive security survey of your home and, at that same time, an estimate of cost. Naturally, there's no obligation!

If you decide on XYZ protection for *both* your professional offices and your home, we do offer a *special pricing plan* for such dual installations.

The attached literature tells a little about XYZ. We would be pleased to tell and show you more. Let us give you an idea of how affordable XYZ can be for both your professional offices and your home. Return the attached reply card, or phone me now, and I'd be happy to arrange an appointment to meet with you.

Sincerely,

(Your Name)
(Your Title)

Encls.

P.S. You may spend thousands of dollars a year on life insurance, yet even more basic is the living insurance and peace of mind that an alarm system can give you in the office and at home. Too many of us say, "It couldn't happen to me," and forget that burglars and fire don't respect income level or status.

14

Advertising for Leads

The Road to Nowhere and Nothing

The preceding chapters were written to *make* money for you. This chapter was written to *save* you money. This chapter is addressed to those who own/manage alarm sales companies and to those who are contemplating entering the business in this mode. Of course, by saving you money, we'll also be helping you to make money, since the money saved and the capital preserved can then be channeled into more productive directions.

When we talk about advertising here, we're including the following:

Television,	Direct mail,
Radio,	Yellow Pages,
Newspaper,	Advertising specialties,
Magazine,	Transit,
Billboards,	Cable.

And while it's always dangerous to generalize, I'm going to give you some advice that will save you a fortune over the course of a career in the alarm business.

When considering the use of any of these media to attempt to generate leads (appointments to make presentations), DON'T. I repeat: *Don't!* None of these media works in any consistent and/or cost-effective way.

Think of it in this light: Your overhead goes on constantly—seven days each week, four weeks each month, and twelve months each year. Your outlays for rent, electricity, payroll, telephone, gasoline, supplies, etc. are all constant, ongoing, and *predictable.*

If history is a criterion (and it is), the only thing that's not predictable in your business is *sales.* It's almost a death sentence for a business, isn't it? Costs are ongoing, but sales are sporadic. Outlays are consistent, but income is intermittent. Expenses are predictable, but closings are merely guesswork. Those factors, of course, are the reasons for the almost 33-1/3% mortality rate every year of burglar alarm sales companies.

Notice the occasional appearance of a burglar alarm sales company's commercials on television. They stay for a few days or a few weeks, and then we don't see them anymore. Why? Talk to some of these alarm companies, and you'll hear horror stories of the large amounts of money expended to generate a few *sporadic* leads. Talk to them about the few leads that ended up costing them $100 apiece. And remember, if you put these leads into the hands of a one-out-of-three closer, you require three leads to make one sale. That's a $300 lead cost. And there aren't even enough TV leads at $100 each.

Notice your Sunday newspaper's TV guide section. Notice the appearance, week in and week out, of dozens of different kinds of home improvement ads: swimming pool ads, keep-your-basement-dry ads, and dozens of other kinds of ads, but no burglar alarm ads. Oh, yes, every once in a while an alarm company ad will appear—for a week, or two weeks—and then it disappears. The reason? Simple: they don't work!

Newspaper? I've met alarm company owners who have bought *full pages* in major newspapers (including 6" x 6" coupons) and received three leads. It doesn't work!

Direct Mail? I've never, in all my years in the alarm sales

business, traveling the entire United States from coast to coast, ever seen a consistent, ongoing, productive direct mail lead-producing program. It is totally non-cost-effective. In fact, very few leads can be generated, even at exorbitant cost.

Again, remember, if we close one out of three, we need fifteen leads to make five sales. If you have three salespeople, you need forty-five leads per week in order to make fifteen sales (five each). And we need these leads every day of every week. Unless we're content to sit around and wait for a sale or two to materialize—maybe! This is the road to averageness at best, bankruptcy at worst.

The Yellow Pages? Those few inquiries that originate from the local Yellow Pages are probably the lowest quality leads imaginable. They are guaranteed to come from prospects who are shopping; thus, they are, in almost every case, calling not only your alarm company but also any number of others.

This means that you are almost invariably involved in a bidding situation, which further means that a one-call close is impossible. It means that you are probably talking to prospects who are interested in the lowest possible price, and this ensures that you will be in a competitive situation where give-away price is everything.

The fact of the matter is that few alarm companies "track" their inquiries precisely. They really don't know what each inquiry from a medium such as the Yellow Pages costs them. The Yellow Pages are expensive. I have met alarm company owners who have spent as much as $50,000 per year in the Yellow Pages. I've *never* met one who could tell me what *each* inquiry cost.

No matter what *you* personally are spending in the Yellow Pages, I implore you to begin tracking your inquiries and then relating these to costs. I guarantee you that, if you do this, you will reduce your Yellow Pages ad(s) to the smallest possible listing.

I suggest also that you track the closing ratio of presentations made to Yellow Pages leads versus other kinds of leads. I know you will find the closing ratio lower for Yellow Page leads and the selling prices lower.

Besides being expensive (expensive would be acceptable if it were cost-effective), Yellow Pages ads are, like the other media mentioned earlier, passive; that is, you pay your money and sit around and hope and wait for an inquiry or inquiries. They may even come once in a while, sporadically. This might be acceptable if your overhead were sporadic. But we know that is not the case. Can you afford to sit around passively and wait for prospects (or even suspects) to come to you? This is why most alarm companies are tiny: The owners are all sitting around waiting for the phone to ring.

My serious advice to you is to learn how to say no to the TV or radio time salesperson, the newspaper space salesperson, and the Yellow Pages salesperson. You must also say no to those selling ballpoint pens with your name on them; ads in the American Legion's annual dinner dance program; ads on bowling alley score sheets; ads on buses, trolleys, billboards; key chains and calendars with your name on them; or *any* other form of advertising. None of these works!

The only thing they do is waste precious dollars, dollars that, if properly used, can generate the constant flow of leads you need every day of every week. By taking my advice, advice based on years of experience and expertise, you will save serious amounts of money. Should you choose to find out for yourself if what I'm saying is true, you'll ultimately come to the same conclusions. The only difference will be that you will be thousands of dollars poorer. Not to mention the fact that your lead flow (the fuel that runs your money machine) will be a sporadic trickle. And that means you can't do the really consistent, big business that is out there to be done.

At this point, you're saying to yourself, "I've just read about all the things I *shouldn't* do. How about what I *should* do?" I'll tell you what you should do—*must* do—in the following chapter. I say "must do," and I say this unequivocally, just as I've told you, in as strong a way as I know how, not to waste your money on advertising.

To follow my advice, with no exceptions, takes discipline. But, then, so does everything else that goes into making a professional. It takes discipline to *be quiet* after asking a closing question. It takes discipline to *walk away* from a presentation where one or the other of the spouses is not present. It takes discipline to attempt a *fifth* trial close. It takes discipline to follow an organized sales presentation *every* single time.

Ask any surgeon whether it's brains or discipline that makes him good at what he does, and he will tell you that it's 99% discipline. That's all that counts. Have the discipline to implement what we tell you in the next chapter, and you will have the key to a constant flow of cost-effective leads forever.

15

The Telephone, the Ultimate Lead Generator

Reach Out and Sell Someone

At the risk of becoming abrasively repetitious (it's worth the risk because it's so critically important), we must underscore again the need for leads on a consistent, predictable basis as well as on a cost-effective basis. There is only *one* way to achieve such an objective, and that is by savvy use of the telephone.

Effective use of the telephone, combined with referrals, cloverleafing (cold canvasing of all premises within a several-block radius of each sale), and individual salespersons' aggressive personal prospecting can produce the number of leads necessary to keep your salespeople in front of enough prospects to make the sales figures impressive (going on the assumption that their closing ratios are at some acceptable level—namely, one out of three or thereabouts). The size of your operation will determine the size of your telephone operation.

I've observed effective, productive telephone operations ranging from one part-time telephone person (producing leads for a one-man enterprise) up to fifteen-plus telephone people producing leads for a twelve-to-fifteen-person sales force.

A rough rule of thumb is one telephone person for each salesperson (if you want to keep that salesperson busy every day). A $5-per-hour telephone person should be able to generate approximately three appointments in a four-hour shift. If this level can be attained, you can reach a $20 lead cost ($5 x 4 hours) per sale. (Remember, you will need three appointments to make one sale.) Even if your telephone person requires eight hours to generate three appointments, you'll still have a viable lead cost: $40 ($5 per hour x 8 hours).

A telephone person should also be paid a commission on each *sale* (not appointment) made from one of their appointments. Paying a commission on a sale is an incentive to make good appointments. A commission in the 1% area is a suggested rate.

Good telephone people *can* be found. With telemarketing becoming one of the fastest growing marketing tools in the business world, experienced telephone people are becoming more common. Thousands of telephone people are employed by some of our must sophisticated marketers such as Sears, Roebuck; J.C. Penney; and others. You can do it, too! And build a business on it.

Don't foolishly look down your nose at the telephone as a powerful lead generator. Don't be as foolish as most of your fellow alarm salespeople. These people resist the telephone because:

- They can't understand how to use it.
- They're afraid of it.
- They personalize; that is, they don't like to receive calls themselves, so they don't believe it works.
- They've tried the telephone in nonprofessional ways and so can't believe anyone else can do it effectively.
- They personally can't handle the rejection that comes over the phone (in large doses) and so can't believe anyone else could handle it either.

What most alarm company owners/managers fail to realize is that good telephone people (people who *like* to work on the telephone) don't treat the rejection personally. They've adjusted their minds to "the numbers game," the game they must play to win.

Let's talk about how you can get started in the telephone game. First, you must develop a script—the actual words for your telephone people to say. We'll help you here, by giving you an example of an effective, tested, tried-and-true telephone script (at the end of this chapter).

Notice something about this script. It's short and it has nothing to do with selling an alarm system. Remember, a basic truism of all telephone activity is *you're not selling a burglar alarm, you're selling an appointment.* Nobody ever bought a burglar alarm over the phone. All they can buy is an appointment. They can give you or your salesperson the opportunity to make an organized sales presentation.

Once you've refined your telephone script, the question becomes "Whom do you call?" You should go to your local public library and look in the section where telephone and street directories are kept. Here, you will find a variety of reverse directories—that is, directories that are arranged by street address rather than by name. Find those directories published covering your area, and from these you will be able to learn who the publishers are and can thus order these directories for yourself.

Ask the reference librarian for *Cole's Directory, Hane's Directory* or R.L. *Polk's Directory.* In some areas, the telephone company publishes reverse street directories. Not all of the directories are published for all areas, but there is a directory of some kind for *your* area.

Some directories even give estimates of the income levels of the households they list. This permits you to call into selected areas where the income levels are high enough so that the prospects can afford to pay for an alarm system, if you're good enough to motivate them to make an affirmative decision. Once you've obtained the directory/directories, you can photocopy those pages you select for calling and assign those to your phone person(s).

I've known alarm company owners or managers who took

the trouble to drive throughout their cities, actually eyeballing various neighborhoods to see which looked to be of some acceptable economic level and then going back to their offices and marking off those neighborhoods in their directories for later calling by their phone people.

Now, the calling begins. You will be making calls by the hundreds—ultimately, by the thousands. And you'll simply be playing the numbers game every hour, every day. You'll hone your script. You'll refine it. You'll test slight changes if it doesn't seem to be producing adequately.

You'll monitor your phone person(s) to determine that:

They're making enough calls per hour;

They're handling the script properly;

They're friendly, warm, and sincere;

They're not making appointments that really aren't appointments;

They're making notes and keeping records of their calls.

You will, no doubt, from time to time have to replace phone people. You may have to hire and fire many before you get the right one(s). It's a constant monitoring and weeding process, no different than with a sales force. *Treat it not as a problem but as a prodecure,* as a challenge and an opportunity.

Don't be discouraged if your initial telephone activities are not instantly successful. Remember, it's a learning process for you, too. It's a sophisticated set of procedures wherein *all* the tumblers have to be in place before it can be fruitful and productive.

Here's a suggestion for you: If you're having trepidation about starting a phone operation, seek out a home improvement company in your area that runs a phone operation. Call and visit the owner and tell him your story. In most cases, because you're not a competitor, such a person will be glad to give you some insights into the running of a phone operation. The home improvement industry is well known for its many good phone operations. There are probably dozens of good phone operations in your area, in a multitude of different businesses. You can learn from all of them.

I implore you to do everything that I've advised and then some! Make yourself "telephone savvy" and you'll eventually find yourself so far ahead of your competition that you'll wonder if they're in the same business! Remember, depending on advertising is the road to no place and nothing. The telephone is the road to consistent lead generation, consistent sales, and thus, serious money.

Following is a very basic telephone script for your telephone people. Notice we don't lie or avoid the issue of why we want to come into their homes; we want to demonstrate our alarm system. To dream up some devious reason (research, surveys, etc.) would only, I guarantee, create hostility in your prospects when they realized the real purpose of your visit.

You may get fewer appointments with this approach, but the people who *do* see you will have *some* degree of legitimate interest. This kind of appointment will result in considerably less rejection for you and/or your salespeople. They may be fewer, but the closing ratio will be higher.

BASIC TELEPHONE PRESENTATION

"Mr. Jones, this is Edna Smith from Atlas Security. How are you today? Mr. Jones, we're the people who are installing so many burglar alarms out in (mention neighborhood) because of the increase of crime in your area. We're going to be in your neighborhood in the next week, Mr. Jones, installing alarms for some of your neighbors, and we would like to make an appointment to give you a demonstration of our new burglar alarm system—at no obligation to you, of course. Would *time 1* (suggest any evening) or *time 2* be more convenient for you?" (Prospect picks a time.)

"OK, Mr. Jones, would you be good enough to write that down? Do you have a pencil handy?" (Repeat the time and date slowly. Then give your name, Atlas Security, and your phone number.)

Prospect's Replies:	*How You Should Respond:*
"I'm not interested."	"I know you're not interested in wasting time, but I assure you that finding out about a system that may save your family's lives wouldn't be a waste of time. And there's no obligation." (Continue with prospect.)
"What's it all about? What are you selling?"	"Basically, Mr. Jones, we manufacture a burglar alarm that can protect you against robbery and fire and that can be installed with a minimum of wiring." (Go on with presentation.)
"What does it cost?"	"Mr. Jones, unless you see the system and learn how it can protect your family and your possessions, giving you a price doesn't make a lot of sense. Except that a burglar alarm system is like a car. You can buy an alarm system with security and lots of options for $2,000 or $3,000. Or you can buy a basic system for $299 to $599." (Go on with presentation; ask for an appointment.) If they say, "That's too much money," you say that it is not too much if it can save a life. If they still resist, thank them and hang up the phone.
"I don't have any money. I can't afford it!"	"Mr. Jones, we would be happy to arrange a financing plan for you for as low as $1.60 a day. I'm sure

Prospect's Replies:	*How You Should Respond:*
	you'll agree that almost anyone can afford $1.60 a day in order to protect their family and possessions." (Go on and ask for an appointment day and time, as previously suggested. Make a note of the money objection on your appointment telephone sheet.
"Who else are you installing it for?"	"Mr. Jones, we maintain a confidential relationship with our clients, in that we, like a lawyer or a doctor, will not give out any information about our clients. We'll respect your privacy in the same way. You wouldn't want to be on a long list of names and get all sorts of calls in the middle of the night, people asking you how you protect your home. I'm sure that makes sense to you, doesn't it?" (Go on and attempt to make the appointment.)
"I don't need to get a pencil to write it down. I have a perfect memory."	"Please just take a second and jot it down. If my boss gets there and if for any reason you forgot the appointment, he'll kill me!"
"Send me some literature."	"We have (nine) systems and (fifty) options. Without surveying your house, all I'll be doing is wasting your time and our postage. It's been our experience that the most intelligent way to pursue this is to

Prospect's Replies:	*How You Should Respond:*
	give you a no-obligation demonstration and a security survey of your home. If you would still like to have some literature, we will leave you the most appropriate literature, for your particular need, at that time."

16

The Anxiety Index

Paranoia Is a Sometime Thing

Some of the people are paranoid all of the time. All of the people are paranoid some of the time. These are not bromides of clichés; they are truisms, derived from a theory I developed many years ago, a theory which, over the years, I've proven to be not really a *theory* but a collection of true *facts.* These facts, when understood, can form the basis for fortune building in the security systems business.

I call this theory—"concept" would be a more accurate label—*the theory of intermittent paranoia.* What it really says is that people have varying degrees of fear, anxiety, paranoia concerning their security and safety. These range from total unconcern to paralyzing fear and all shades in between.

They undergo these different stages of fear at different times and to different degrees. I visualize it as an "anxiety index," graduated from one to ten:

10 Terrified zone: "When can you put it in?"
9 Three people on the block were burglarized: "When can we see it?"

8 Lots of crime in the area: "We probably should take a hard look!"

7 "We've been giving it consideration."

6 "It's been somewhere in the back of our minds.

5 }
4 } The "I don't need it!" or no-money zone.
3 }
2 }
1 }

As we've said several times earlier, "You can't sell them a solution to a problem they don't think they have." So, we're foolish to waste our precious time on those people who fall into the one-through-five gradation on our anxiety index. Their fear coefficient is too low. In fact, because it's so low, they probably won't even give anyone an opportunity to make a presentation to them.

We want to concentrate our efforts and time on the only people who are logical prospects: the sixes through tens on the anxiety index. These people, to varying degrees, are fearful enough at least to listen to our presentation. And a six or seven can, via an effective presentation, be elevated into the nine or ten level and then motivated to make a positive decision.

Of course, much of the time, people who have been fours on the anxiety index become nines or tens overnight. All it takes is for their next-door neighbor to be burglarized or for them to discover jimmy marks on their back door upon returning from the movies some night.

They become terrified. They vow to take some action to protect themselves. They become nines. And then, in most cases, they do nothing. They procrastinate. And, in a few days, or a few weeks, they revert to being fours on the anxiety index. Shortly thereafter, someone is mugged three blocks from their home. And they become sevens or eights again on the anxiety index. Once again, they proceed to do nothing and fall back to being fives or sixes.

Their paranoia, you see, is intermittent; it comes and goes. Threes become nines and fours soar to the ten level based on occurrences they experience or that are experienced by friends, neighbors, or relatives. Or maybe they just read or hear about some frightening happening.

Because of the workings of intermittent paranoia, the stream of citizens who become likely prospects is never ending. The same person who very nearly hangs up on your telephone person one day will, on another day, ask how soon your company can have someone come out to discuss a security system.

The catch, of course, is that it is strictly coincidence when your telephone person calls someone at the very time they are riding high on the anxiety index. The happenstance may be that your call is made at a low ebb on the anxiety index, and you make no repeat call two months later when the same person is considerably higher on the index because of some fear-producing occurrence. This tells us several things. It gives us insights into telephone prospecting that we must have if we are to truly understand the way to make it work.

This first and most basic principle we learn is that because paranoia (and the resultant receptivity to your message) is intermittent, the chances that your phone call is being made to a prospect at a high point on his or her anxiety index are purely and simply coincidence. The only way we can combat this luck factor is to make large numbers of calls. It's only in this way that we can encounter enough prospects who are receptive enough to allow us to set up an appointment for a presentation.

It's the old (but oh-so-true) bromide of all selling activities, the numbers game, at work again! The more phone calls made, the better the chances of generating more appointments. (Remember, that's what we're selling to begin with: appointments.)

Once we understand this principle, we then concern ourselves with the only thing that counts: the volume of phone calls made per hour. We call into play the most powerful law of the universe. It's not Murphy's Law, it's not Ohm's Law, it's not Newton's Law. No! It's the Law of Averages. It sounds so simple, almost simplistic, and yet it's so little understood by fledglings (and sometimes even veteran) alarm sales company operators. Live by this law and you will become wealthy.

Another principle we learn from the concept of intermittent paranoia is that your telephone lists (names to be called) are never dead. After going through each list and having called all those on that list, the list should be put into a file and used again in two months (or three, if you have enough names). The point is that certain people on each list, who were low on the anxiety index on one calling, will be much higher on the index on a call back. Remember that people keep going up and down, and then up again, on the anxiety index. It goes on forever. Your lists are good forever.

Proof of the validity of the intermittent paranoia concept is easy to come by. It is really before our eyes constantly. Think along with me, if you will. Periodically, everyone in the alarm business gets phone calls from people who say they want an alarm system, and could someone come out to show them a system and discuss its immediate installation.

In many cases, when we visit these people we find that they've lived in their homes for many years with no alarm system. What possessed these people to call us at the specific time they did, with the urgency they had? Why now and not before, during all those previous years? The answer is obvious. Something occurred that took them from years of being low on the anxiety index to being tens. They perceive themselves in danger now, whereas they didn't have this perception before. We're fortunate in being in a profession through which we can help them. Help them realize that *paranoia is nothing more than extreme realism.*

17

A Philosophy of Protection

The Hardware Syndrome

A home without a burglar alarm is an invitation to a personal tragedy. You've really got to believe this statement with all your heart and mind. This is critical. If you don't have this conviction, how can you expect the prospect to have it?

After all, closing is intimately related to *conviction.* You, by means of an organized sales presentation, have imbued your prospect(s) with the conviction that your alarm system will solve their problem. Your prospects' conviction is a rub-off of your own conviction—the knowledge/belief that your alarm system will do the job. If you don't radiate this conviction, can you logically expect the prospect(s) to reflect it? Of course not.

One of the questions that arises now is Yes, but what alarm system? Which one can I believe in? Which one can I sell with conviction? We're not here to launch into a lengthy technical discussion of alarm system hardware. What's more important is to talk about a philosophy of protection.

Our primary goal must be to get as many people, families, and homes protected as is in our power to do. Obviously, we'll make money only in proportion to how successful we are in getting the numbers. The more families protected, the more commissions and/or profits.

This goes almost without saying, but another aspect of our endeavors is almost as crucial: Every family that we succeed in protecting is one family that's not a candidate for becoming a crime statistic and a personal tragedy. We must *believe* in this credo. Live by it! Doing so will pay large dividends for both ourselves and for society. But we'll never get very many families protected if our philosophy of protection demands that we protect every home as if it were the U.S. Mint or Fort Knox.

In my constant travels throughout the country, visiting local and regional alarm associations, I am constantly amazed that the bulk of the content of their programs and events are centered around hardware—new widgets, gadgets, and space age technology. Rarely does one visit one of these conclaves without witnessing a seemingly endless series of discussions, seminars, and workshops devoted to alarm hardware, ever more complicated and expensive.

Rarely does one hear the single most critical problem of the industry addressed: how to get the 98% of the people, who are unprotected, protected? The vast majority of the populace simply can't afford the kinds of expensive, elaborate alarm systems the alarm industry has traditionally attempted to sell them. Consequently, most people (even many tens on the anxiety index) do nothing to protect themselves, purely for economic reasons.

We hear interminable discussions of the merits of passive infrared versus ultrasonic, microwave versus redundant space protection, but little or nothing about how to sell large numbers of affordable alarm systems to citizens who desperately need them. One practically never hears a group of alarm people discussing professional selling techniques or marketing programs. The talk is always of nuts and bolts. In a way, of course, this is a subconsciously defensive posture. Since most alarm people don't have the foggiest notion about professional selling techniques, they hide among the widgets.

Because of this hardware syndrome, the alarm business has remained almost a cottage industry in the face of the greatest crime wave the world has ever known, and a great number of people have gone unprotected. Recently, I was in Houston, Texas, for a speaking date. On my arrival, I took a taxi from the airport to my hotel, and on the highway from the airport into the city, I was struck by an interesting phenomenon. I saw electric signs in front of automobile dealers' showrooms saying "Buick-Honda Dealer," "Pontiac-Subaru Dealer," and "Cadillac-Chevrolet-Datsun Dealer"—every conceivable variation. What suddenly dawned on me was the fact that the automobile industry had finally awakened to the idea that there wasn't just an automobile market, there were automobile markets. And so, rather than have people leave their showrooms without buying anything, the automobile dealers realized they had to offer something for every pocketbook, every budget.

Thus, now the savvy auto dealer is able to sell the affluent prospect a Buick and the budget-conscious young couple a Subaru. But at least they now have a shot at selling transportation to a broad spectrum of the prospects they are able to generate. The point here is that you, the alarm professional, must be able to offer a full perimeter alarm system to the affluent prospect with the expensive home but, at the same time, be able to offer a basic, economy system to the newlyweds in a tract house.

As you saw in our organized sales presentation, one of the most critical areas is the place where we ask the prospect what price level he would be comfortable with, what his budget can stand. What is the point of showing the prospect your new eight-zone control panel and your space age system for $2,500 when the very stretched limit of his budget is $850. You'll end up with no sale and no profit or commission, and the prospect will end up minus the protection we all know he needs. *Some* protection is better than *no* protection at all.

Don't treat the medium-priced suburban home as if it were a high-security-risk fur vault. Tailor a basic, economical system they can afford, and feel good about getting them *some* protection. Generally, the system you sell them will be quite adequate to thwart the amateur intruders who make up the vast majority of attempted/completed burglaries.

You should have multiple pricing levels, along with multiple levels of protection, no different than Sears, Roebuck, who offers almost all major, big-ticket appliances in ranges of good-better-best, providing an affordable price level for all types of budgets. Sears doesn't let 'em get away because of price.

There are dozens of manufacturers of excellent alarm components; thus, putting in a quality alarm system, whatever the scope of it, is not difficult. What's difficult is restraining yourself from trying to make every system you install similar to a bank system. As we discussed in Chapter 9, you must avoid the hardware peddlar's syndrome. If you *do* avoid it, you'll end up selling lots more hardware.

I chuckle ironically to myself frequently when I hear an alarm company operator bemoaning the feared inroads expected to be made by the cable TV companies onto the alarm industry's turf. The irony of this situation is that, for the most part, the cable TV industry has been almost totally ineffective in selling volume quantities of alarm systems. The reason? They became infected with the hardware peddlar's syndrome and proceeded to emulate the conventional alarm industry. The cable industry has, and is, making all of the same profound philosophical mistakes that have been part and parcel of the alarm business for decades.

There is no need for any concern about the cable ogre. Cable is not the enemy; *we* are the enemy. With astronauts actually walking around on the moon and traffic jams evolving among space shuttles, the conventional alarm companies are still snaking wires through walls.

Alarm companies (the mainstream) continue to be alarm *installers* rather than alarm *sellers.* Don't pride yourself on being a proficient alarm installer; they get paid like mechanics. Professional alarm *sellers* get paid like doctors.

Don't pride yourself on the expensive, elaborate system you sold to the tycoon in the big house on the hill. You'll deposit more dollars in your bank account by selling *dozens* of systems to the middle Americans on the other side of town. Actually, if your philosophy of protection is smart, you'll sell 'em all!

Forget the conventional wisdom because it isn't very wise. It's the same so-called wisdom that managed to leave 98% of the homes in the United States unprotected (and unsold). And while your colleagues and competitors are sitting around at the alarm association meeting discussing false alarms and telephone rates, you (and your people) will be out making presentations, motivating prospects to say yes, and *closing*.

Contrary to what its title would lead you to believe, this book is *not* about the burglar alarm business. What it's really about is changing your life and about fortune building. Whatever "rich" means to you, I promise you that if you do all of the difficult things I've advised you to do, and if you make a part of you the philosophies I've espoused, you will get there. I promise you.

~~THE END~~

THE BEGINNING